Recent Advances in Mathematical Sciences

Adem Kılıçman · Hari M. Srivastava
M. Mursaleen · Zanariah Abdul Majid
Editors

Recent Advances in Mathematical Sciences

Selected Papers from ICREM7 2015

Springer

Editors
Adem Kılıçman
Universiti Putra Malaysia
Serdang
Malaysia

M. Mursaleen
Aligarh Muslim University
Aligarh
India

Hari M. Srivastava
University of Victoria
Victoria
Canada

Zanariah Abdul Majid
Universiti Putra Malaysia
Serdang
Malaysia

ISBN 978-981-10-0517-6 ISBN 978-981-10-0519-0 (eBook)
DOI 10.1007/978-981-10-0519-0

Library of Congress Control Number: 2016931321

Printed on acid-free paper

This Springer imprint is published by SpringerNature
The registered company is Springer Science+Business Media Singapore Pte Ltd.

Preface

The International Conference on Research and Education in Mathematics (ICREM) is a biennial event which aims to bring together academicians, scientists, and industrialists from around the country and the world for knowledge sharing, exchanging ideas, and for collaborating and presenting research results about all aspects of mathematics and its applications. The first four ICREM were held in Kuala Lumpur in the years 2003, 2005, 2007, and 2009. Following this, the conference was successfully held in Bandung, Indonesia (2011) and in Ho Chi Minh City, Vietnam (2013).

In conjunction with that, the Institute for Mathematical Research (INSPEM), Universiti Putra Malaysia took the responsibility to organize the 7th International Conference on Research and Education in Mathematics (ICREM7) with technically co-sponsored by IEEE Malaysia WIE Affinity Group. This conference took place on August 25–27, 2015 in Kuala Lumpur, Malaysia. The main goal of ICREM7 is to provide opportunities for the delegates to exchange new ideas and application experiences, to establish research or business relations, and to find global partners for future collaboration. The conference also included several invited papers on important and timely topics from well-known leaders in the field, and parallel tracks of oral presentation sessions of the accepted papers.

This book contains research papers by invited speakers and selected papers from the fields of applied mathematics and pure mathematics in ICREM7. In chapter "Counting Trees and Rooted Trees with Applications," our keynote address for this conference was given by Emeritus Prof. Dr. John Butcher, the mathematician specialized in numerical analysis. The principal applications discussed in this talk are connected with the structures of Runge–Kutta methods and canonical Runge–Kutta methods. Another eight chapters in this book represent a cross section of state-of-the-art and cutting-edge research in pure mathematics and applied mathematics for future.

We thank all the following supporting agencies and individuals for their contribution and significant assistance for ICREM7:

- IEEE Malaysia WIE Affinity Group
- Akademi Sains Malaysia (ASM)
- Bandung Institute of Technology (ITB)
- Institute for Mathematics, Vietnam Academy of Science and Technology (IMVAST)
- Malaysian Mathematical Sciences Society (PERSAMA)
- OEMS IntiPakar Corporation Sdn Bhd
- The Indonesian Mathematical Society (IndoMS)
- Putrajaya Corporation
- Corporate Strategy and Communications Office (CosComm), UPM
- Nur Raidah Salim
- Mohamed Faris Laham
- Kathiresan Gopal
- Phang Pei See
- Zahari Mahad

The editors

Contents

Counting Trees and Rooted Trees with Applications

J.C. Butcher

Abstract Trees are connected graphs with no cycles. Rooted trees have a specific vertex designated to be the root. The order of a tree is the number of vertices. As the order increases the total number of trees or rooted trees with this order grows rapidly. A generating function for these totals will be demonstrated. The principal applications discussed in this talk are connected with the structures of Runge–Kutta methods and canonical Runge–Kutta methods.

Keywords Trees · Rooted trees · Generating functions · Runge–Kutta methods · Order conditions

Mathematics Subject Classification (2000) 65L05

1 Introduction

We will discuss counting rooted trees

$$= \{\mathbf{t}_1, \quad \mathbf{t}_2, \quad \mathbf{t}_3, \quad \mathbf{t}_4, \quad \mathbf{t}_5, \quad \mathbf{t}_6, \quad \mathbf{t}_7, \quad \mathbf{t}_8, \quad \dots\}$$

J.C. Butcher (✉)
The University of Auckland, Auckland, New Zealand
e-mail: butcher@math.auckland.ac.nz

A. Kılıçman et al. (eds.), *Recent Advances in Mathematical Sciences*,
DOI 10.1007/978-981-10-0519-0_1

and unrooted trees

$$U = \{\ \bullet,\ \ldots\ \} = \{\mathbf{u}_1,\ \mathbf{u}_2,\ \mathbf{u}_3,\ \mathbf{u}_4,\ \mathbf{u}_5,\ \ldots\}$$

The order of a rooted tree $|t|$, or unrooted tree $|u|$, is the number of vertices. For example

$$\begin{aligned} &|\mathbf{t}_1| = 1,\ |\mathbf{t}_2| = 2,\ |\mathbf{t}_3| = |\mathbf{t}_4| = 3,\ |\mathbf{t}_5| = |\mathbf{t}_6| = |\mathbf{t}_7| = |\mathbf{t}_8| = 4,\\ &|\mathbf{u}_1| = 1,\ |\mathbf{u}_2| = 2,\qquad |\mathbf{u}_3| = 3,\qquad\qquad |\mathbf{u}_4| = |\mathbf{u}_5| = 4.\end{aligned}$$

Some unrooted trees have a special type of symmetry and these are called superfluous trees (S-trees). These are formed by joining two identical rooted trees together at the roots. For example, joining two copies of the rooted tree on the left gives the unrooted superfluous tree on the right

The set of S-trees will be denoted by S. We see that only those S-trees with order not exceeding 4 are $\mathbf{u}_2$ and $\mathbf{u}_5$.

Define three generating functions

$$\mathcal{A}(x) = \sum_{n=1}^{\infty} a_n x^n, \text{ where } a_n \text{ is the number of members of } T \text{ with order } n.$$

$$\mathcal{B}(x) = \sum_{n=1}^{\infty} b_n x^n, \text{ where } b_n \text{ is the number of members of } U \text{ with order } n.$$

$$\mathcal{C}(x) = \sum_{n=1}^{\infty} c_n x^n, \text{ where } c_n \text{ is the number of members of } U \text{ with order } n, \text{ which are not also members of } S.$$

Just by looking at the first few rooted and unrooted trees, and counting, we observe that

$$\begin{aligned} \mathcal{A}(x) &= x + x^2 + 2x^3 + 4x^4 + \cdots\\ \mathcal{B}(x) &= x + x^2 + x^3 + 2x^4 + \cdots\\ \mathcal{C}(x) &= x \qquad\ \ + x^3 + \ x^4 + \cdots \end{aligned}$$

The first aim of this paper is to present formulae for $\mathcal{A}(x)$, $\mathcal{B}(x)$, and $\mathcal{C}(x)$. The result for $\mathcal{A}(x)$ is a classical result in combinatorics, whereas for $\mathcal{B}(x)$ and $\mathcal{C}(x)$ the principal reference will be [5]. The derivation of the generating functions, presented here, makes use of the tree and forest spaces which are introduced in Sect. 2. The applications are to the order conditions for Runge–Kutta methods and for symplectic Runge–Kutta methods.

2 The Tree and Forest Spaces

A forest is a formal product of trees. The set of all forests will be named F. The empty product will be denoted by 1.

If f is a forest then $[f]$ is the tree formed by joining all the factors in f to a new root. As an example, the forest $\mathbf{t}_1^2\mathbf{t}_3\mathbf{t}_6$ (on the left) and the tree $[\mathbf{t}_1^2\mathbf{t}_3\mathbf{t}_6]$ (on the right) are:

Lemma 1 *For every $t \in T$ there is a unique forest f such that $t = [f]$.*

Proof Remove the root of t. If nothing is left, then the unique forest is $f = 1$. Otherwise f is the collection of trees that remains.

The tree space is the vector space of formal linear combinations of trees. The forest space F is the vector space of formal linear combinations of forests. Products are found by formal multiplication as in the example:

$$\begin{aligned}\left(1 + 2\mathbf{t}_1\mathbf{t}_2\right)\left(3\mathbf{t}_1 + 4\mathbf{t}_1^2\,\mathbf{t}_2\right) &= 3\mathbf{t}_1 + 4\mathbf{t}_1^2\mathbf{t}_2 + 6\mathbf{t}_1\mathbf{t}_2\mathbf{t}_1 + 8\mathbf{t}_1\mathbf{t}_2\mathbf{t}_1\mathbf{t}_1\,\mathbf{t}_2\\ &= 3\mathbf{t}_1 + 10\mathbf{t}_1^2\,\mathbf{t}_2 + 8\mathbf{t}_1^3\,\mathbf{t}_2^2\,.\end{aligned}$$

Define the vector space X as the set of linear combinations of powers of x. Let $\boldsymbol{\phi}$ be the mapping from F to X, where 1 is mapped to 1 and every tree t is mapped to $x^{|t|}$. For example,

$$\begin{aligned}\boldsymbol{\phi}\big((1 + 2\mathbf{t}_1\mathbf{t}_2)(3\mathbf{t}_1 + 4\mathbf{t}_1^2\,\mathbf{t}_2)\big) &= \boldsymbol{\phi}(1 + 2\mathbf{t}_1\mathbf{t}_2)\boldsymbol{\phi}(3\mathbf{t}_1 + 4\mathbf{t}_1^2\,\mathbf{t}_2)\\ &= (1 + 2x^3)(3x + 4x^4) = 3x + 10x^4 + 8x^7.\end{aligned}$$

3 Counting Rooted Trees

An interesting member of $\mathbf{F}$ is the sum of all trees: We can calculate ϕ of this:

$$\begin{aligned}\phi\left(\sum_{t\in T} t\right) &= |\mathbf{t}_1| + |\mathbf{t}_2| + |\mathbf{t}_3| + |\mathbf{t}_4| + \cdots \\ &= x + x^2 + x^3 + x^3 + \cdots \\ &= x + x^2 + 2x^3 + 4x^4 + 9x^5 + \cdots \\ &= \mathcal{A}(x).\end{aligned}$$

Another interesting member of $\mathbf{F}$ is the sum of all powers of a single tree t

$$\begin{aligned}\sum_{n=0}^{\infty} t^n &= 1 + t + t^2 + t^3 + t^4 + \cdots \\ &= (1-t)^{-1},\end{aligned}$$

with

$$\begin{aligned}\phi\left(\sum_{n=0}^{\infty} t^n\right) &= 1 + |t| + |t|^2 + |t|^3 + |t|^4 + \cdots \\ &= (1-|t|)^{-1}\end{aligned}$$

Even more interesting is the product over all trees

$$\prod_{t\in T}(1-t)^{-1}$$

because this gives the sum over all forests with

$$\phi\left(\sum_{\mathbf{F}\in\mathbf{F}} \mathbf{F}\right) = \prod_{t\in T}(1-x^{|t|})^{-1} = \prod_{n=1}^{\infty}(1-x^n)^{-a_n}.$$

Theorem 1 *The value of a_n is the coefficient of x^{n-1} in*

$$\prod_{m=1}^{n-1}(1-x^m)^{-a_m}.$$

Proof For every $t \in T$ there is a unique forest f such that $t = [f]$

$$\sum_{t\in T} t = \left[\prod_{t\in T} (1-t)^{-1} \right].$$

4 Counting Unrooted Trees

If t_1 and t_2 are rooted trees then the (noncommutative) product $t_1 \circ t_2$, also known as the "Butcher product", is defined as a rooted tree with order $|t_1| + |t_2|$ formed by joining their roots. The root of t_1 becomes the root of the new rooted tree as in the diagram

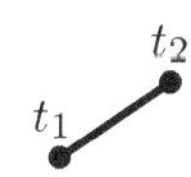

Unrooted trees can be thought of as equivalent classes of rooted trees, with $t_1 \circ t_2 \equiv t_2 \circ t_1$ a generator of the equivalence relation. The tree containing both $t_1 \circ t_2$ and $t_2 \circ t_1$ is

Among the members of T up to order 4, we have

$$\begin{aligned}
\mathbf{t}_3 = \mathbf{t}_2 \circ \mathbf{t}_1 &\equiv \mathbf{t}_2 \circ \mathbf{t}_2 = \mathbf{t}_4 \\
\mathbf{t}_5 = \mathbf{t}_3 \circ \mathbf{t}_1 &\equiv \mathbf{t}_3 \circ \mathbf{t}_2 = \mathbf{t}_7 \\
\mathbf{t}_6 = \mathbf{t}_4 \circ \mathbf{t}_1 &\equiv \mathbf{t}_1 \circ \mathbf{t}_4 = \mathbf{t}_8,
\end{aligned}$$

so that we can complete the identification, to this order, of unrooted trees with equivalent classes of rooted trees:

$$\begin{aligned}
\mathbf{u}_1 &= \{\mathbf{t}_1\}, \\
\mathbf{u}_2 &= \{\mathbf{t}_2\}, \\
\mathbf{u}_3 &= \{\mathbf{t}_3, \mathbf{t}_4\}, \\
\mathbf{u}_4 &= \{\mathbf{t}_5, \mathbf{t}_7\}, \\
\mathbf{u}_5 &= \{\mathbf{t}_6, \mathbf{t}_8\}.
\end{aligned}$$

A vertex v is a "centre" of an unrooted tree u if the rooted tree with root v is $[t_1 t_2 \dots t_m]$, with

$$\max_{i=1}^{m} |t_i| \le \tfrac{1}{2}|t|$$

If $|t|$ is even, it is possible to have bicenters; that is two adjacent vertices each of which is a center.

Theorem 2 (Sanz-Serna, Abia)

$$\mathcal{B}(x) = \mathcal{A}(x) - \tfrac{1}{2}(\mathcal{A}(x)^2 - \mathcal{A}(x^2)), \tag{1}$$
$$\mathcal{C}(x) = \mathcal{A}(x) - \tfrac{1}{2}(\mathcal{A}(x)^2 + \mathcal{A}(x^2)) \tag{2}$$

Proof To enumerate unrooted trees, we count only trees which are central or bicentral. In the latter case only one of the bicentral pair is counted and we note that for S-trees, the members of the bicentral pair are identical. We carry out this enumeration by counting all trees and subtracting the ones which do not qualify.

As an example of the use of generating functions, calculate $\mathcal{A}(x)$ up to degree 10. This is found to be

$$\begin{aligned}\mathcal{A}(x) &= x + x^2 + 2x^3 + 4x^4 + 9x^5 + 20x^6 + 48x^7 + 115x^8 + 286x^9 + 719x^{10} + \cdots \\ &= x(1-x)^{-1}(1-x^2)^{-1}(1-x^3)^{-2}(1-x^4)^{-4}(1-x^5)^{-9}\cdots\end{aligned}$$

To calculate $\mathcal{B}(x)$ and $\mathcal{C}(x)$, we also need

$$\begin{aligned}\mathcal{A}(x)^2 &= x^2 + 2x^3 + 5x^4 + 12x^5 + 30x^6 + 74x^7 + 188x^8 + 478x^9 + 1235x^{10} + \cdots \\ \mathcal{A}(x^2) &= x^2 + x^4 + 2x^6 + 4x^8 + 9x^{10} + \cdots\end{aligned}$$

Substitute into (1) and (2) and we obtain the results

$$\begin{aligned}\mathcal{B}(x) &= x + x^2 + x^3 + 2x^4 + 3x^5 + 6x^6 + 11x^7 + 23x^8 + 47x^9 + 106x^{10} + \cdots \\ \mathcal{C}(x) &= x + x^3 + x^4 + 3x^5 + 4x^6 + 11x^7 + 19x^8 + 47x^9 + 97x^{10} + \cdots\end{aligned}$$

5 Order of Runge–Kutta Methods

A Runge–Kutta method is characterized by a tableau

$$\begin{array}{c|c} c & A \\ \hline & b^{\mathsf{T}} \end{array} = \begin{array}{c|cccc} c_1 & a_{11} & a_{12} & \cdots & a_{1s} \\ c_2 & a_{21} & a_{22} & \cdots & a_{2s} \\ \vdots & \vdots & \vdots & & \vdots \\ c_s & a_{s1} & a_{s2} & \cdots & a_{ss} \\ \hline & b_1 & b_2 & \cdots & b_s \end{array},$$

where $c = A\mathbf{1}$.

Given an initial value problem

$$y'(x) = f(y(x)), \quad y(x_0) = y_0, \quad \text{where} \quad y_0 \in \mathbb{R}^N, \quad f : \mathbb{R}^N \to \mathbb{R}^N, \qquad (3)$$

step number n of the numerical solution is computed, along with stages Y_i, $i = 1, 2, \ldots, s$ from the equations

$$Y_i = y_{n-1} + h \sum_{j=1}^{s} a_{ij} f(Y_j), \quad i = 1, 2, \ldots, s,$$

$$y_n = y_{n-1} + h \sum_{i=1}^{s} b_i f(Y_i).$$

The numerical order of the method is determined by conditions on the coefficients in the tableau. These conditions take the form $\Phi(t) = 1/t!$, where for each tree t, $\Phi(t)$ is a polynomial of degree $|t|$, in the coefficients appearing in the method tableau, and the factorial of the tree $t!$, is an integer depending on the particular tree. The condition must hold whenever $|t| \le p$.

The meanings of $\Phi(t)$ and $t!$ are illustrated in the examples

t	$\Phi(t) = \frac{1}{t!}$
$\mathbf{t}_1$	$b^{\mathsf{T}}\mathbf{1} = 1$
$\mathbf{t}_2$	$b^{\mathsf{T}}c = \frac{1}{2}$
$\mathbf{t}_3$	$b^{\mathsf{T}}c^2 = \frac{1}{3}$
$\mathbf{t}_4$	$b^{\mathsf{T}}Ac = \frac{1}{6}$
$\mathbf{t}_5$	$b^{\mathsf{T}}c^3 = \frac{1}{4}$
$\mathbf{t}_6$	$b^{\mathsf{T}}cAc = \frac{1}{8}$
$\mathbf{t}_7$	$b^{\mathsf{T}}Ac^2 = \frac{1}{12}$
$\mathbf{t}_8$	$b^{\mathsf{T}}A^2c = \frac{1}{24}$

For a more complicated example, using a tree t with order 10. The expression for $\Phi(t)$ is calculated from the labels attached to the vertices on the left and $t!$ is calculated from the integers attached on the right,

The results are found to be

$$\Phi(t) = \sum_{i,j,k,\ell,m=1}^{s} b_i a_{ij} c_j a_{ik} c_k a_{k\ell} c_\ell a_{\ell m} c_m^2,$$
$$t! = 3 \cdot 5 \cdot 7 \cdot 2 \cdot 10 = 2100.$$

Theorem 3 *A Runge–Kutta method has order p if and only if*

$$\Phi(t) = \frac{1}{t!}$$

for all t such that $|t| \le p$.

This result is proved in [1] and in [2] and [3].

6 Canonical Runge–Kutta Methods

A method (A, b^{T}, c) is canonical (or symplectic) if

$$\mathrm{diag}(b)A + A^{\mathsf{T}}\,\mathrm{diag}(b) = bb^{\mathsf{T}}.$$

When these methods are applied to Hamiltonian problems, they preserve symplectic behaviour, see [4]. They also preserve the quadratic invariant $\|y(x)\|^2$ in a problem satisfying

$$\langle Y, f(Y)\rangle = 0, \quad \forall Y. \tag{4}$$

We have

Theorem 4 *If f in (3) satisfies (4) and $y_0, y_1, \ldots$ are computed by a canonical Runge–Kutta method, then*

$$\|y_n\| = \|y_{n-1}\| = \|y_0\|, \quad n = 1, 2, \ldots.$$

Proof We have

$$\begin{aligned}
&\langle y_n, y_n\rangle - \langle y_{n-1}, y_{n-1}\rangle \\
&\quad = \langle y_{n-1} + h\sum_{i=1}^{s} b_i f(Y_i), y_{n-1} + h\sum_{j=1}^{s} b_j f(Y_i)\rangle - \langle y_{n-1}, y_{n-1}\rangle \\
&\quad = h\sum_{i=1}^{s} b_i \langle f(Y_i), y_{n-1}\rangle + h\sum_{j=1}^{s} b_j \langle y_{n-1}, f(Y_j)\rangle + h^2 \sum_{i,j=1}^{s} b_i b_j \langle f(Y_i), f(Y_j)\rangle
\end{aligned}$$

Substitute $b_i b_j = b_i a_{ij} + b_j a_{ji}$ and the result can be written as

$$h\sum_{i=1}^{s} b_i\langle f(Y_i), Y_i\rangle + h\sum_{j=1}^{s} b_j\langle Y_j, f(Y_j)\rangle = 0.$$

This result applies also to any quadratic form in place of $\langle\cdot,\cdot\rangle$. However, its most important application is to Hamiltonian problems, [4] and Canonical methods are also referred to as "symplectic" methods.

To illustrate the advantages of canonical methods, three attempts have been made to solve the Harmonic oscillator problem

$$\begin{aligned} y_1' &= -y_2, \qquad & y_1(0) &= 1,\\ y_2' &= y_1, \qquad & y_2(0) &= 0, \end{aligned}$$

where we note that $\|y(x)\| = 1$ is an invariant for this problem.

In each attempt, a Runge–Kutta method was used with $h = 2\pi/40$ and 2000 steps were taken; this is equivalent to approximately 50 orbits. The three methods were (E) an explicit second-order method, (I) an implicit second-order method, and (M) the implicit midpoint rule method, with tableaux

$$E:\ \begin{array}{c|cc} 0 & 0 & 0\\ \frac12 & \frac12 & 0\\ \hline & 0 & 1 \end{array}, \qquad I:\ \begin{array}{c|cc} \frac12 & 1 & -\frac12\\ 1 & 1 & 0\\ \hline & 1 & 0 \end{array}, \qquad M:\ \begin{array}{c|c} \frac12 & \frac12\\ \hline & 1 \end{array}.$$

The results are shown in Fig. 1 where the advantages of the symplectic method M are clearly observed. The path traced out by this method stays on the unit circle, forever.

It was shown in [5] that unrooted trees, as well as rooted trees, have a significant role in the order analysis in the case of canonical methods.

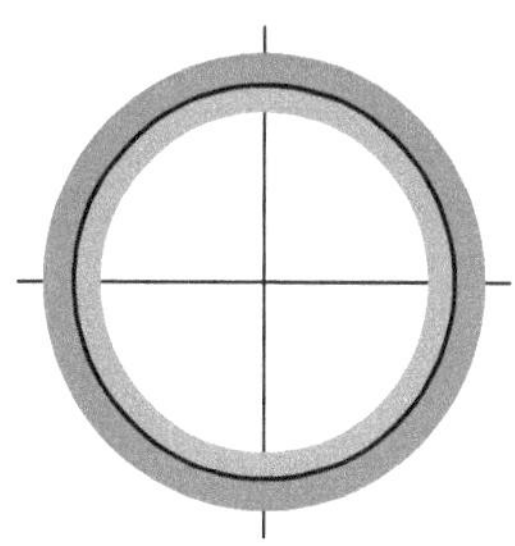

Fig. 1 Numerical results for harmonic oscillator for the methods E (*dark grey*), I (*light grey*) and M (*black line*)

7 Order Conditions for Canonical Methods

We want to find the relationship between the order conditions associated with the equivalent trees $t_1 \circ t_2$ and $t_2 \circ t_1$. The conclusion will be that for canonical methods these conditions imply each other and one of them can be omitted. Furthermore, in the superfluous case in which $t_1 = t_2$, the order condition associated with $t_1 \circ t_2$ can be disregarded entirely.

Write the order conditions for t_1 and t_2 in the forms

$$\Phi(t_1) = b^{\mathsf{T}}\Psi(t_1) = \frac{1}{t_1!},$$
$$\Phi(t_2) = b^{\mathsf{T}}\Psi(t_2) = \frac{1}{t_2!},$$

where $\Psi(t_1), \Psi(t_2)$ are vectors dependent on A and c but not b^{T}, then we can evaluate $\Phi(t_1 \circ t_2)$ and $\Phi(t_2 \circ t_1)$

$$\Phi(t_1 \circ t_2) = b^{\mathsf{T}}\big(\Psi(t_1) \circ A\Psi(t_2)\big) = \Psi(t_1)^{\mathsf{T}}\big(\operatorname{diag}(b)A\big)\Psi(t_2),$$
$$\Phi(t_2 \circ t_1) = b^{\mathsf{T}}\big(\Psi(t_2) \circ A\Psi(t_1)\big) = \Psi(t_1)^{\mathsf{T}}\big(A^{\mathsf{T}}\operatorname{diag}(b)\big)\Psi(t_2).$$

This leads to the result

$$\begin{aligned}\Phi(t_1 \circ t_2) + \Phi(t_2 \circ t_1) &= \Psi(t_1)^{\mathsf{T}}\big(\operatorname{diag}(b)A + A^{\mathsf{T}}\operatorname{diag}(b)\big)\Psi(t_2)\\ &= \Psi(t_1)^{\mathsf{T}}\big(bb^{\mathsf{T}}\big)\Psi(t_2)\\ &= \Phi(t_1)\Phi(t_2) \qquad (5)\end{aligned}$$

We can also calculate

$$\frac{1}{(t_1 \circ t_2)!} + \frac{1}{(t_2 \circ t_1)!} = \frac{1}{t_1!t_2!}. \qquad (6)$$

These results lead to a key theorem.

Theorem 5 (Sanz-Serna, Abia) *Given trees t_1 and t_2, where $|t_1| + |t_2| = p$ then for a method with order $p-1$,*

$$\Phi(t_1 \circ t_2) = \frac{1}{(t_1 \circ t_2)!} \quad \textit{if and only if} \quad \Phi(t_2 \circ t_1) = \frac{1}{(t_2 \circ t_1)!}.$$

Furthermore, if $t_1 = t_2$, then

$$\Phi(t_1 \circ t_2) = \frac{1}{(t_1 \circ t_2)!}.$$

Proof From (5) and (6) we deduce

$$\left(\Phi(t_1 \circ t_2) - \frac{1}{(t_1 \circ t_2)!}\right) + \left(\Phi(t_2 \circ t_1) - \frac{1}{(t_2 \circ t_1)!}\right) = \Phi(t_1)\Phi(t_2) - \frac{1}{t_1!t_2!} = 0$$

and the result follows.

This means in effect that, for a canonical method, rather than having to impose an order condition of each rooted tree up to the required order, it is only necessary to satisfy a single-order condition for each nonsuperfluous (nonrooted) tree.

8 Conclusions

Both rooted and unrooted trees have fundamental roles in the theory of Runge–Kutta methods in that the number of order p conditions for a Runge–Kutta method is the number of rooted trees with order not exceeding p. In the case of symplectic or canonical methods, the number of independent conditions is the number of unrooted trees to order p, where superfluous trees are omitted. In this survey paper, enumerations for rooted trees and unrooted trees are given in terms of generating functions and the connection with the order of numerical methods is shown.

References

1. Butcher, J.C.: Coefficients for the study of Runge–Kutta integration processes. J. Aust. Math. Soc. **3**, 185–201 (1963)
2. Butcher, J.C.: Numerical Methods for Ordinary Differential Equations. Wiley, Hoboken (2008)
3. Hairer, E., Nørsett, S.P., Wanner, G.: Solving Ordinary Differential Equations I: Nonstiff Problems. Springer, Berlin (1993)
4. Sanz-Serna, J.M.: Runge–Kutta schemes for Hamiltonian systems. BIT **39**, 877–883 (1988)
5. Sanz-Serna, J.M., Abia, L.: Order conditions for canonical RK schemes. SIAM J. Numer. Anal. **28**, 1081–1096 (1991)

Boundedness and Stability of Leslie–Gower Model with Sokol–Howell Functional Response

Safaa Jawad Ali, Norihan Md. Arifin, Raid Kamel Naji, Fudziah Ismail and Norfifah Bachok

Abstract In this chapter, a three-species model of Leslie–Gower predator–prey food chain model with Sokol–Howell functional response is proposed. The boundedness of the solution of the model is discussed. Local and global stability analyses of the system are carried out. The dynamics of the predator–prey food chain model with Sokol–Howell functional response is investigated theoretically as well as numerically.

Keywords Food chain · Chaotic · Leslie–Gower · Functional response · Sokol–Howell

1 Introduction

The work of May [1] exploring the chaotic behaviors of population dynamics inspired much research work in the predator–prey system [2–11]. Alaoui [12] proposed and studied the dynamics of a modified Leslie–Gower predator–prey food chain model with Holling type II functional response. Naji et al. [5] studied a modified Leslie–Gower food chain model with Bendigton–DeAnglis functional response and the

S.J. Ali (✉) · N.Md. Arifin · F. Ismail · N. Bachok
Institute for Mathematical Research, Universiti Putra Malaysia, 43400 Serdang, Malaysia
e-mail: safa_ali1962@yahoo.com

N.Md. Arifin
e-mail: norihana@upm.edu.my

F. Ismail
e-mail: fudziah@upm.edu.my

N. Bachok
e-mail: norfifah@upm.edu.my

S.J. Ali
Technical Preparing Institute, Middle Technical University, Baghdad, Iraq

R.K. Naji
Department of Mathematics, College of Science, University of Baghdad, Baghdad, Iraq
e-mail: rknaji@gmail.com

A. Kılıçman et al. (eds.), *Recent Advances in Mathematical Sciences*,
DOI 10.1007/978-981-10-0519-0_2

model exhibited chaotic dynamics. Gakkhar with Priyadarshi [13] studied the Leslie–Gower food web system. The numerical works of Alaoui, Naji, and Gakkhar are brilliant and perfect.

In their experiments about the kinetics of phenol oxidation, Sokol and Howell [14] suggested a simplified Holling type IV function of the form $\frac{wx}{d+x^2}$ and found that it is simpler and better than the original function of Holling type IV. The Holling type IV response represents a situation in which the predation of the predator decreases at sufficiently high prey densities [10] and about how this functional response is obtained, see [15–17]. Ruan [18] and Hu [19] both studied the dynamics and bifurcation analysis of continuous-time and discrete-time models of modified Holling type IV, respectively. Investigations on Leslie–Gower type model [12, 13, 20, 21] and Sokol–Howell functional response [14, 18, 19, 22, 23] are relatively less than the other types like Lotka–Volterra and Bendigton–DeAnglis functional responses.

This paper is organized as follows: in Sect. 2, the mathematical model is proposed and each parameter in the model is described. In Sect. 3, the boundedness of the solution of the model is established. Stability analyses of the equilibrium points of the model are derived in Sect. 4. In Sect. 5, numerical study is carried out to obtain the behavior of the model. Finally, the paper ends with a conclusion in Sect. 6.

2 The Mathematical Model

Consider the three-species food chain model at time t consisting of the prey population density denoted by $x(t)$, the middle predator population density denoted by $y(t)$, and the top predator whose population density denoted by $z(t)$. The middle predator y preys on its sole food x at the lower level according to simplified Holling type IV functional response, while the top predator z preys on y at the second level according to the modified Leslie–Gower type.

The dynamics of the model described above can be represented by the following set of differential equations:

$$
\begin{aligned}
\frac{dx}{dt} &= x\,(a_0 - b_0 x) - \frac{v_0 xy}{d_0 + x^2} : \quad x(0) \geq (0),\\
\frac{dy}{dt} &= \frac{v_1 xy}{d_1 + x^2} - a_1 y - \frac{v_2 yz}{d_2 + y} : \quad y(0) \geq (0),\\
\frac{dz}{dt} &= c_3 z^2 - \frac{v_3 z^2}{d_3 + y} : \quad z(0) \geq 0.
\end{aligned} \tag{1}
$$

Here the positive constants a_0, b_0, v_0, d_0, v_1, a_1, d_1, v_2, d_2, c_3, v_3, and d_3 denote: a_0 is the growth rate of the prey x, b_0 represents the intraspecific competition among individuals of prey x, v_i's are the maximum values attainable by each per capita rate, d_0 and d_1 measure the extent to which the environment provides protection to the prey x and predator y, respectively, a_1 represent the death rate of y in the absence of x, d_2 is the value of y at which the per capita removal rate of y becomes $\frac{v_2}{2}$, c_3

represents the growth rate of z by sexual reproduction, the number of males and females being assumed to be equal, d_3 represents the residual loss in z population due to serve scarcity of its favorite food y; the second term on the right-hand side in the third equation of system (1) depicts the loss in the top predator population.

Remark: The origin of the model of system (1) is standard in first two equations, but the third equation is absolutely not standard. About how the third equation obtained, see [12, 13, 20, 21].

Now according to the third equation of system (1):

$$\frac{dz}{dt} = z\left(c_3 z - \frac{v_3 z}{d_3 + y}\right),$$

if the middle predator y is absence ($y = 0$), the top predator goes extinct if

$$c_3 d_3 < v_3, \tag{2}$$

and increase without bound if $c_3 d_3 > v_3$. In this paper, we will suppose that condition (2) holds.

Then, the system (1) when $d_0 = d_1$ can be written as follows:

$$\begin{aligned} \frac{dx}{dt} &= x\left(a_0 - b_0 x - \frac{v_0 y}{d_1 + x^2}\right) = G_1(x, y, z), \\ \frac{dy}{dt} &= y\left(\frac{v_1 x}{d_1 + x^2} - a_1 - \frac{v_2 z}{d_2 + y}\right) = G_2(x, y, z), \\ \frac{dz}{dt} &= z\left(c_3 z - \frac{v_3 z}{d_3 + y}\right) = G_3(x, y, z), \end{aligned} \tag{3}$$

with $x(0) \geq 0$, $y(0) \geq 0$, and $z(0) \geq 0$. Obviously, the interaction functions G_i ($i = 1, 2, 3$) of system (3) are continuous and have continuous partial derivatives on the positive octant $R^3_+ = \{(x, y, z) \in R^3_+ : x \geq 0, y \geq 0, z \geq 0\}$. Therefore, these functions are Lipschitzian on R^3_+, and hence the solution of the system (3) exists and is unique.

3 Boundedness of the Model

In this section, the boundedness of the solution of the system (3) in R^3_+ is established in the next theorem.

Theorem 1 *All the solutions of the three-species food chain system (3) are uniformly bounded, provided*

$$\frac{a_0 v_1}{b_0 v_0} + \frac{a_0^2 v_1}{4 a_1 b_0 v_0} + d_3 < \frac{v_3}{c_3}, \tag{4}$$

and let Ω be the set defined by

$$\Omega = \left\{(x, y, z) \in R_+^3 : 0 \leq x \leq \frac{a_0}{b_0},\ 0 \leq x + \frac{v_0}{v_1} y \leq \frac{a_0}{b_0} + \frac{a_0^2}{4a_1 b_0},\right.$$
$$\left. 0 \leq x + \frac{v_0}{v_1} y + \alpha z \leq \frac{a_0}{b_0} + \frac{a_0^2}{4a_1 b_0} + \frac{N}{a_1}\right\}$$

where

$$\alpha = \frac{1}{a_1^2\left(\frac{a_0 v_1}{b_0 v_0} + \frac{a_0^2 v_1}{4a_1 b_0 v_0} + d_3\right)} \tag{5}$$

$$N = \frac{1}{4\left(v_3 - \left(\frac{a_0 v_1}{b_0 v_0} + \frac{a_0^2 v_1}{4a_1 b_0 v_0} + d_3\right) c_3\right)} \tag{6}$$

Proof Let $(x(t), y(t), z(t))$ be any solution of the system with non-negative initial condition. Now there are three cases about the boundedness of the solutions.

- Case 1: To prove that x(t) is bounded $\forall t \geq 0$.

Since we have

$$\frac{dx}{dt} \leq x\left(a_0 - b_0 x\right), \tag{7}$$

then according to comparison theorem [24], we obtain that

$$\lim_{t\to\infty} Sup\ x(t) \leq \frac{a_0}{b_0} \tag{8}$$

implies that $x(t) \leq \frac{a_0}{b_0}$, $\forall t \geq 0$. Now as $t \to \infty$.

$$\Rightarrow 0 \leq x(t) \leq \frac{a_0}{b_0}.$$

- Case 2: To prove that x(t) and y(t) are bounded $\forall t \geq 0$.

Consider

$$M_1(t) = x(t) + \frac{v_0 y(t)}{v_1}, \quad M_1(0) \geq 0$$

Then

$$\frac{dM_1(t)}{dt} \leq x(a_0 - b_0 x) - \frac{a_1 v_0}{v_1} y$$

Since in $\Omega, 0 \leq x \leq \frac{a_0}{b_0}$ and simplification using $Max_{[0,\frac{a_0}{b_0}]}x(a_0 - b_0 x) = \frac{a_0^2}{4b_0}$ gives

$$\frac{dM_1(t)}{dt} + a_1 M_1(t) \leq \frac{a_0}{b_0} + \frac{a_0^2}{4a_1 b_0} \tag{9}$$

Therefore for all $t \geq 0$

$$M_1(t) \leq \left(\frac{a_0}{b_0} + \frac{a_0^2}{4a_1 b_0}\right) - \left[\left(\frac{a_0}{b_0} + \frac{a_0^2}{4a_1 b_0}\right) - M_1(0)\right] e^{-a_1 M_1 t} \tag{10}$$

Hence as $t \to \infty$, since $(x(0), y(0), z(0)) \in \Omega$

$$x(t) + \frac{v_0}{v_1} y(t) \leq \frac{a_0}{b_0} + \frac{a_0^2}{4a_1 b_0} \quad \forall t \geq 0. \tag{11}$$

Similarly for Case 3

$$0 \leq x + \frac{v_0}{v_1} y + \alpha z \leq \frac{a_0}{b_0} + \frac{a_0^2}{4a_1 b_0} + \frac{N}{a_1}.$$

Therefore, every solution initiated in non-negative octant are attracted in a bounded set Ω defined above, which implies to the uniformly bounded of $y(t)$ and $z(t)$. Thus the proof is complete.

4 Stability of the Model

There are at most three non-negative equilibrium points of system (3) in addition to the positive equilibrium point E_3 in $R+^3$ existence and stability conditions of them are given as follows:

- The trivial equilibrium point $E_0 = (0, 0, 0)$ always exits.
- The equilibrium point $E_1 = (\frac{a_0}{b_0}, 0, 0)$ always exists on the boundary of the first octant.
- The middle predator can exist and survive depending on its prey. Therefore, the equilibrium point $E_2 = (\bar{x}, \bar{y}, 0)$ exists uniquely in the positive quadrant of x–y plane where $\bar{x}$ and $\bar{y}$ are given by:

$$\bar{x} = \frac{v_1}{2a_1}, \quad \bar{y} = \frac{1}{v_0}\left(a_0 - b_0 \frac{v_1}{2a_1}\right)(d_1 + \bar{x}^2), \tag{12}$$

provided that the following conditions hold

$$\frac{v_1}{2a_1} < \frac{a_0}{b_0}, \quad v_1^2 - 4{a_1}^2 d_1 = 0 \tag{13}$$

- In the absence of prey x, then both y and z cannot survive, so there is no equilibrium point in the y–z plane. In addition to that, if the middle predator y is absent, then there is no equilibrium point in the x–z plane.
- The positive equilibrium point $E_3 = (x^*, y^*, z^*)$ exists in the interior of the first octant if and only if there is a positive solution to the following equations:

$$\begin{aligned} f_1 &= a_0 - b_0 x - \frac{v_0 y}{d_1 + x^2} = 0, \\ f_2 &= \frac{v_1 x}{d_1 + x^2} - a_1 - \frac{v_2 z}{d_2 + y} = 0, \\ f_3 &= c_3 z - \frac{v_3 z}{d_3 + y} = 0. \end{aligned} \tag{14}$$

Straightforward computation shows that

$$y^* = \frac{v_3}{c_3} - d_3, \tag{15}$$

while x^* is the positive root of the following equation

$$x^3 - \frac{a_0}{b_0} x^2 + d_1 x + \left[\frac{1}{b_0} \left(v_0 y^* - a_0 d_1 \right) \right] = 0,$$

this equation can be rewritten as

$$f(x) = Ax^3 + Bx^2 + Cx + D = 0, \tag{16}$$

where $A = 1$, $B = -\frac{a_0}{b_0}$, $C = d_1$ and $D = \left[\frac{1}{b_0} \left(v_0 y^* - a_0 d_1 \right) \right]$.

Now since $0 \leq x^* \leq \frac{a_0}{b_0}$, then $f(0) = D < 0$, if

$$y^* < \frac{a_0}{v_0} d_1, \tag{17}$$

$f(\frac{a_0}{b_0}) = \frac{v_0}{b_0} y^* > 0$. Thus, $f(0) f(\frac{a_0}{b_0}) < 0$, and then there is a positive root of Eq. (16) lies in $(0, \frac{a_0}{b_0})$ when $y^* < \frac{a_0}{v_0} d_1$, is satisfied.

The second equation of (14) gives

$$z^* = \frac{(d_2 + y^*)}{v_2} \left(\frac{v_1 x^*}{d_1 + x^{*2}} - a_1 \right) \tag{18}$$

Therefore, the positive equilibrium point $E_3 = (x^*, y^*, z^*)$ exists if in addition to conditions (2) and (17), the following condition holds:

$$a_1 < \frac{v_1 x^*}{(d_1 + x^{*2})}. \tag{19}$$

Now in order to investigate the dynamical behavior of the three species food chain system (3) near the above equilibrium points, the variational matrix V of system (3) at (x, y, z) is computed as:

$$V(x, y, z) = \begin{bmatrix} x\frac{\partial f_1}{\partial x} + f_1 & x\frac{\partial f_1}{\partial y} & x\frac{\partial f_1}{\partial z} \\ y\frac{\partial f_2}{\partial x} & y\frac{\partial f_2}{\partial y} + f_2 & y\frac{\partial f_2}{\partial z} \\ z\frac{\partial f_3}{\partial x} & z\frac{\partial f_3}{\partial y} & z\frac{\partial f_3}{\partial z} + f_3 \end{bmatrix}$$

where $\frac{\partial f_1}{\partial x} = -b_0 + \frac{2v_0xy}{(d_1+x^2)}$, $\frac{\partial f_1}{\partial y} = -\frac{v_0}{(d_1+x^2)}$, $\frac{\partial f_1}{\partial z} = 0$, $\frac{\partial f_2}{\partial x} = \frac{v_1(d_1-x^2)}{(d_1+x^2)}$, $\frac{\partial f_2}{\partial y} = \frac{v_2z}{(d_2+y)^2}$, $\frac{\partial f_2}{\partial z} = -\frac{v_2}{(d_2+y)}$, $\frac{\partial f_3}{\partial x} = 0$, $\frac{\partial f_3}{\partial y} = \frac{v_3z}{(d_3+y)^2}$, $\frac{\partial f_3}{\partial z} = c_3 - \frac{v_3}{d_3+y}$.

Further, the stability analysis of the system (3) is carried out and according to the variational matrix $V_i; i = 0, 1, 2, 3$ of $E_i; i = 0, 1, 2, 3$ respectively, the following results are obtained:

$$V_0 = \begin{bmatrix} a_0 & 0 & 0 \\ 0 & -a_1 & 0 \\ 0 & 0 & 0 \end{bmatrix},$$

$$V_1 = \begin{bmatrix} -a_0 & -\frac{v_0}{d_1+1} & 0 \\ 0 & \frac{v_1}{d_1+1} - a_1 & 0 \\ 0 & 0 & 0 \end{bmatrix},$$

$$V_2 = \begin{bmatrix} \bar{x}\left(-b_0 + \frac{2\bar{x}(a_0-b_0\bar{x})}{(d_1+\bar{x}^2)}\right) & -\frac{v_0\bar{x}}{(d_1+\bar{x}^2)} & 0 \\ \frac{v_1\bar{y}(d_1-\bar{x}^2)}{(d_1+\bar{x}^2)^2} & \frac{v_1\bar{x}}{(d_1+\bar{x}^2)} - a_1 & \frac{-v_2\bar{y}}{d_1+y} \\ 0 & 0 & 0 \end{bmatrix}.$$

- Here the trivial equilibrium point E_0 is a nonhyperbolic saddle-node, having an unstable manifold along x-direction.
- From variational matrix V_1, it is observed that the nonhyperbolic equilibrium point E_1 is a saddle point having stable manifold along x-direction if the following condition holds

$$\frac{v_1}{d_1 + 1} > a_1, \tag{20}$$

  while nonhyperbolic equilibrium point E_1 having stable manifold along x and y-direction if

$$\frac{v_1}{d_1 + 1} < a_1. \tag{21}$$

- The equilibrium point $E_2 = (\bar{x}, \bar{y}, 0)$ is a nonhyperbolic point having stable manifold along x and y-direction if the following conditions hold.

$$\frac{2\bar{x}(a_0 - b_0\bar{x})}{(d_1 + \bar{x}^2)^2} < b_0, \quad \frac{v_1\bar{x}}{d_1 + \bar{x}^2} < a_1 \tag{22}$$

while E_2 is unstable saddle if the opposite of any part of condition (22) hold.

However, for the positive equilibrium point $E_3 = (x^*, y^*, z^*)$, the variational matrix is:

$$V_3 = \begin{bmatrix} a_{11} & a_{12} & a_{13} \\ a_{21} & a_{22} & a_{23} \\ a_{31} & a_{32} & a_{33} \end{bmatrix} \tag{23}$$

The characteristic equation of the variational matrix (23) can be written as

$$\lambda^3 + A_1\lambda^2 + A_2\lambda + A_3 = 0$$

where $A_1 = -(a_{11} + a_{22})$, $A_2 = a_{11}a_{22} - a_{12}a_{21} - a_{23}a_{32}$, and $A_3 = a_{11}a_{23}a_{32}$. According to Routh–Hurwitz criterion, $E_3 = (x^*, y^*, z^*)$ is locally asymptotically stable provided $A_1 > 0$, $A_3 > 0$, and $\Delta = A_1A_2 - A_3 > 0$.

Now straightforward computations show that, $A_1 > 0$ and $A_3 > 0$ if and only if the following conditions are satisfied:

$$a_0 < 2b_0x + \frac{v_0y^*(d_1 - x^{*2}) - v_1x^*R}{R^2} + \left(a_1 + \frac{v_2z^*}{Q_1^2}\right), \tag{24}$$

with

$$\frac{v_1}{v_0} < \frac{y^*(d_1 - x^{*2})}{x^*R}, \quad x^{*2} < d_1. \tag{25}$$

In addition to that, since

$$A_1A_2 - A_3 = (a_{11} + a_{22})(a_{12}a_{21} - a_{11}a_{22}) + a_{22}a_{23}a_{32}.$$

Hence, the necessary condition for $A_1A_2 - A_3 > 0$ is

$$(a_{12}a_{21} - a_{11}a_{22}) < 0,$$

which is equivalent to the following condition

$$\left[a_0 - \left(2b_0x^* + \frac{v_0y^*(d_1 - x^{*2})}{R^2}\right)\right] \left[\frac{v_1x^*}{R} - \left(a_1 + \frac{v_2d_2z^*}{Q_1^2}\right)\right] + \frac{v_0v_1x^*y^*(d_1 - x^{*2})}{R^3} > 0. \tag{26}$$

Therefore, depending on the above analysis, the locally asymptotically stable in $Int R_+^3$ of the positive equilibrium point $E_3 = (x^*, y^*, z^*)$ is discussed in the following theorem

Theorem 2 *Suppose that positive equilibrium point* $E_3 = (x^*, y^*, z^*)$ *exists in* $Int R_+^3$, *then* E_3 *is locally asymptotically stable provided conditions (24–26) hold.*

Proof Follows directly from Routh–Hurwitz criterion [25].

5 Numerical Simulation

In this section, the global dynamics of system (3) are studied numerically. The food chain system is solved numerically using predictor–corrector method with six order Runge–Kutta method [26]. System (3) run for 60,000 time steps and the first 30,000

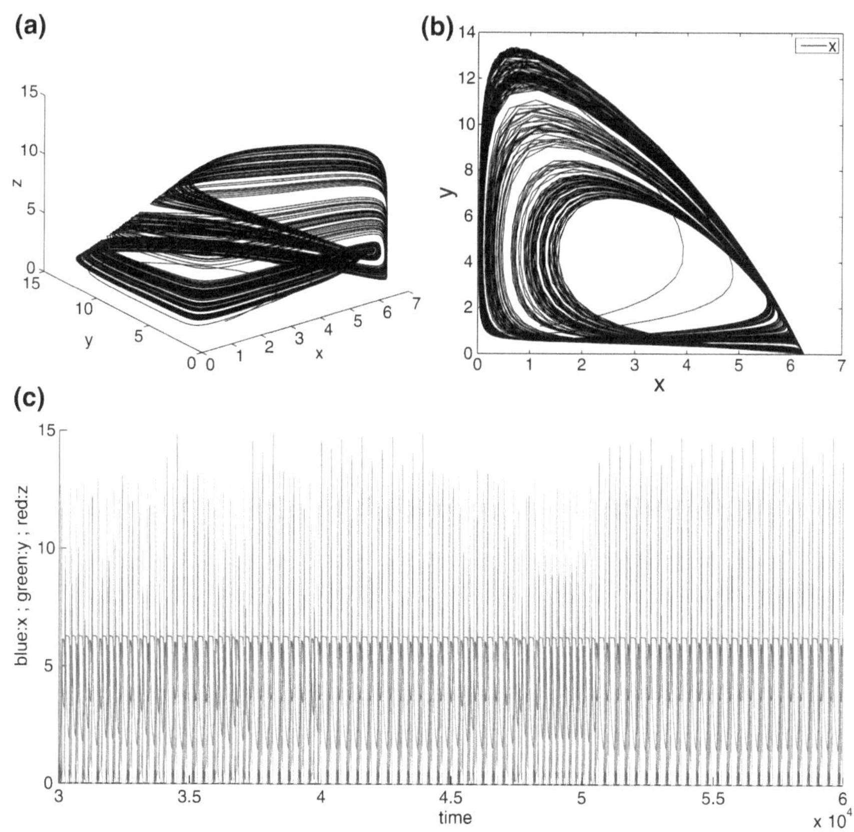

Fig. 1 **a** 3D of system (3) chaotic attractor for data (27) with $a_0 = 0.47$ and $c_3 = 0.047$, **b** 2D x–y plane of Fig. 1a. **c** Time series of Fig. 1a

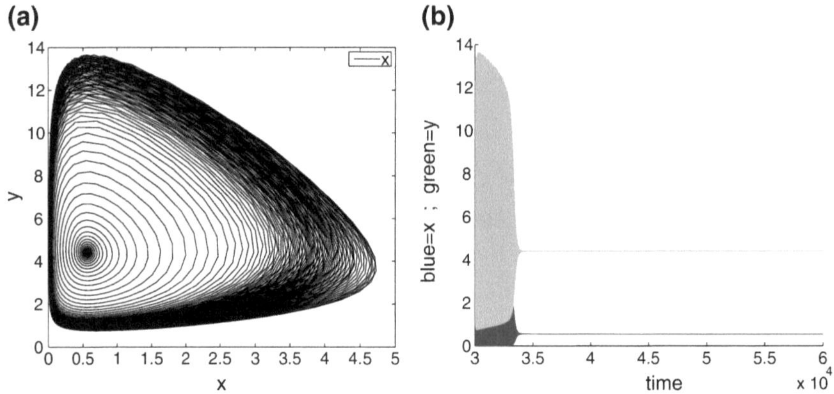

Fig. 2 **a** 2D x–y plane of system (3) asymptotically stable for data (27), $a_0 = 0.47$ and $c_3 = 0.040$, **b** Time series of Fig. 2a

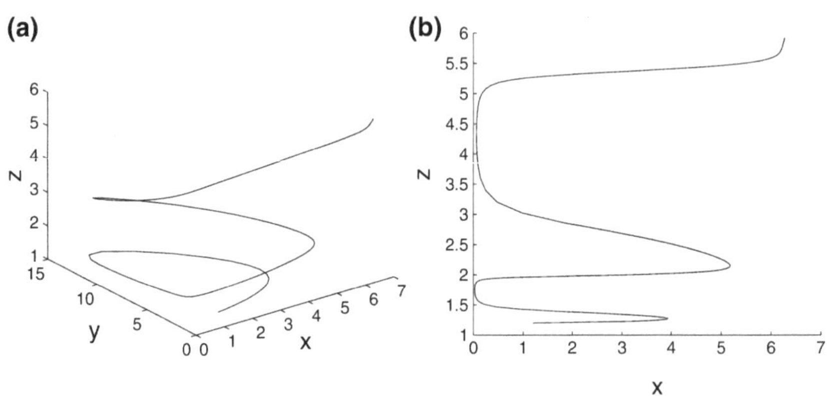

Fig. 3 **a** 3D of system (3) asymptotically stable for data set (27), $a_0 = 0.47$ and $c_3 = 0.050$, **b** 2D x–z plane of Fig. 3a

time steps are deleted to eliminate the transient effect. For the following set of fixed parameter values

$$\begin{aligned} &a_1 = 0.105,\ b_0 = 0.075,\ d_1 = 10.0,\ d_2 = 10.0, \\ &d_3 = 20.0,\ v_0 = 1.0,\ v_1 = 2.0,\ v_2 = 0.405,\ v_3 = 1.0. \end{aligned} \tag{27}$$

The attractor of system (3) in 3D and 2D with their time series are drawn in Fig. 1 for the initial condition (1.2, 1.2, 1.2). The main objective is to explore the possibility of chaotic behavior of system (3) by depending on the controlling parameter and keeping other parameters of (27) fixed.

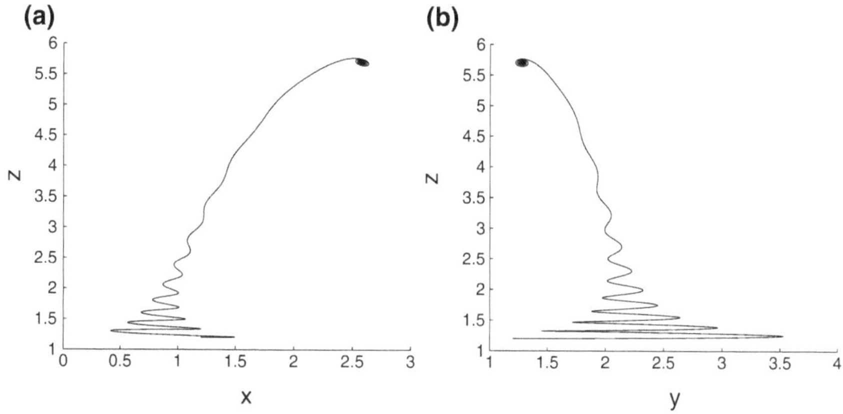

Fig. 4 **a** 2D x–z plane of system (3) asymptotically stable point for data in (27) with $a_0 = 0.27$ and $c_3 = 0.047$, **b** 2D y–z plane of system (3) with same data of Fig. 4a

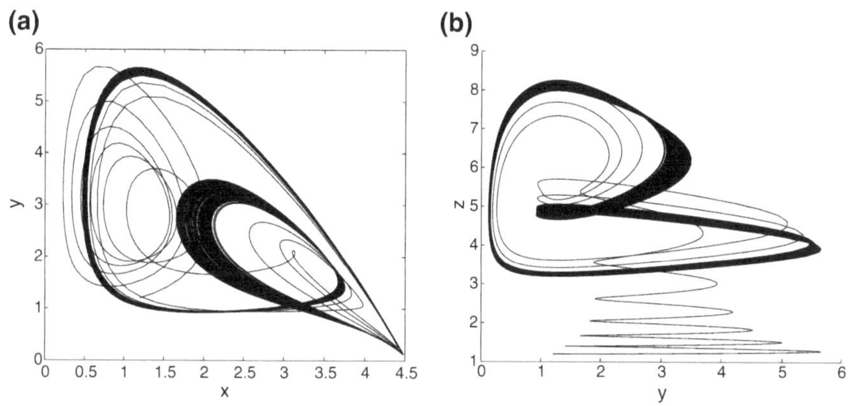

Fig. 5 **a** 2D x–y plane of system (3) period doubling attractor with $a_0 = 0.34$ and $c_3 = 0.047$, **b** 2D y–z plane of Fig. 5a

- The first case by fixing $a_0 = 0.47$ and varying the value of c_3 in the range 0.041–0.049, it is observed that the system approaches to chaotic dynamics of system (3) as shown in Fig. 1 for $c_3 = 0.047$, while decreasing the value of c_3 less than 0.041 stabilizing the system as shown in Fig. 2 for the typical value $c_3 = 0.040$. Further increasing the of c_3 more than 0.049 change the dynamics of system (3) from chaotic to stable point (6.265, 1.003e−192, 5.910) as shown in Fig. 3 for the typical value $c_3 = 0.050$.
- The second case is by fixing $c_3 = 0.047$ and varying the value of a_0 in the range 0.27–0.047 with data in (27). For the value $a_0 = 0.27$ the system approach to stable at the point (2.577, 1.276, 5.700) as in Fig. 4. Increasing the value of $a_0 = 0.34$,

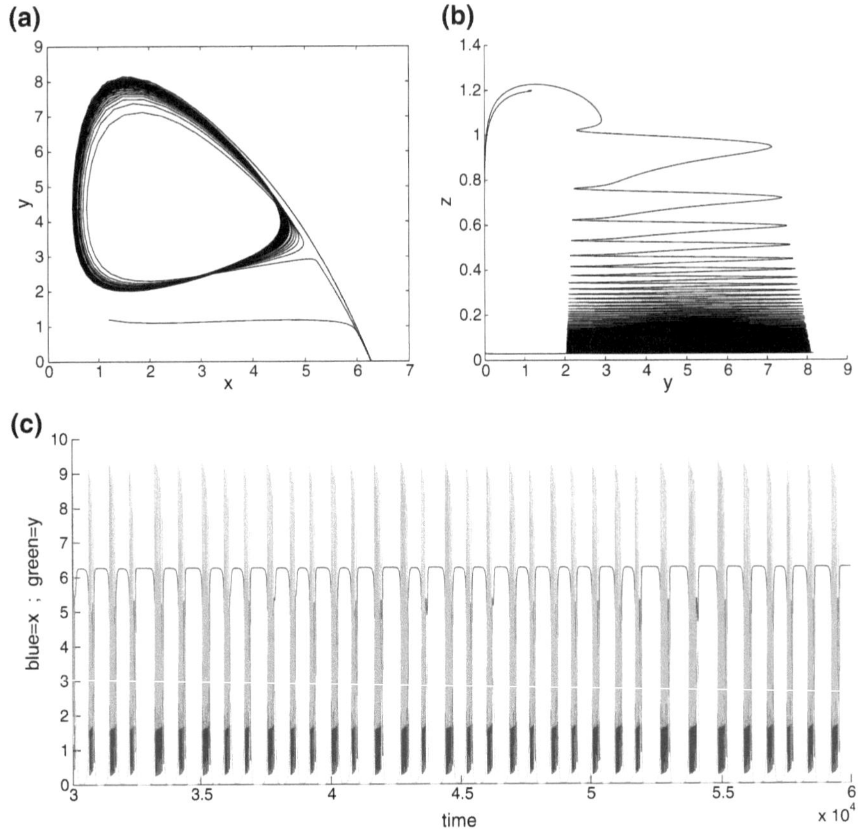

Fig. 6 **a** 2D x–y plane of system (3) strange attractor approach to periodic for data in (27) with $a_0 = 0.47$, $c_3 = 0.047$ and $a_1 = 0.25$, **b** 2D y–z plane tea-cup attractor for data set Fig. 6a, **c** time series of Fig. 6a

change the dynamics of system (3) to period doubling as shown in Fig. 5, while increasing a_0 more than 0.034 change the dynamics of the system to chaotic for the typical value of $a_0 = 0.47$ as shown in Fig. 1.

- The third case about fixing $a_0 = 0.47$ and $c_3 = 0.047$ with same data in (27), it is observed by increasing the death rate of the middle predator $a_1 = 0.25$ then system (3) behavior is chaotic approach to periodic as shown in Fig. 6. Further increasing of a_1 more than 0.25 change system (3) dynamics from chaotic to asymptotically stable.

6 Conclusion

In order to explain the dynamical behavior of the proposed food chain systems (3), local as well as global stability analysis are carried out. Boundedness of the system is discussed.

In addition to that to confirm the analytical results, the system is solved numerically for different sets of biologically feasible parameter values, and then the attracting sets with their time series are drown in order to explain the dynamical behavior of the model as in Figs. 1–6 with data the same as in (27). According to our study, the following results are obtained:

(1) The intrinsic growth rate of the top predator c_3 is a sensitive parameter which lead to sensitivity of system (3) dynamics. Decreasing c_3 less than 0.041 and increasing c_3 more than 0.049 lead to change the dynamics of the food chain model from chaotic to asymptotically stable as shown in Figs. 2 and 3, which prove that a small change in c_3 will lead to major change in the dynamics of system (3).
(2) System (3) has a chaotic dynamics as in Fig. 1, but if we decreasing the values of intrinsic growth of the prey species a_0, then system (3) approaches to periodic attractor as in Fig. 5 and decreasing a_0 more change the behavior of the system to an asymptotically stable as in Fig. 4, so decreasing a_0 has a stabilizing effect on the dynamics of system (3) and a_0 is control parameter of system (3).
(3) Increasing the death rate a_1 in the middle predator y has a stabilizing effect in the dynamics of system (3) as shown in Fig. 6.

Acknowledgments The authors would like to thank the financial support received from Ministry of Higher Education, Malaysia in the form of an FRGS research grant and the first author would like to thank INSPEM.

References

1. May, R.M.: Simple mathematical models with very complicated dynamics. Nature **261**, 459–467 (1976)
2. Sahoo, B., Poria, S.: The chaos and control of a food chain model supplying additional food to top-predator. Chaos Solitons Fractals **58**, 52–64 (2014)
3. Gakkhar, S., Singh, A.: Control and chaos due to additional predator in Hastings-powell food chain model. J. Math. Anal. Appl. **385**, 423–438 (2012)
4. Gakkhar, S., Naji, R.K.: Chaos in three-species ratio dependent food chain. Chaos Solitons Fractals **14**, 771–778 (2002)
5. Naji, R.K., Upadhyay, R.K., Rai, V.: Dynamical consequences of predator interference in tri-trophic model food chain. Nonlinear Anal. Real World Appl. **11**, 809–818 (2010)
6. Gilpin, M.E.: Spiral chaos in predator-prey model. Am. Nat. **107**, 306–308 (1979)
7. Hasting, A., Powell, T.: Chaos in three-species food chain. Ecology **72**, 896–903 (1991)
8. Kuznetsov, Ya., Rinaldi, S.: Remarks on food chain dynamics. Math. Biosci. **134**, 1–33 (1996)
9. Peet, A.B., Deutsch, P.A., Peacock-Lopez, E.: Complex dynamics in a three-level system with intraspecies interaction. J. Theor. Biol. **232**, 491–503 (2005)

10. Upadhyay, R.K., Iyengar, S.R.K.: Introduction to Mathematical Modeling and Chaotic Dynamics, pp. 121–122. CRC Press, A Chapman and Hall Book, Boca Raton (2014)
11. Upadhyay, R.K., Iyengar, S.R.K., Rai, V.: Chaos: an ecological reality. Int. J. Bifurc. Chaos **8**(6), 1325–1333 (1998)
12. Aziz-Aloui, M.A.: Study of a Leslie–Gower type tritrophic population model. Choas Solitons Fractals **14**, 1275–1293 (2002)
13. Leslie, P.H., Gower, J.C.: The properties of stochastic model for the predator-prey type of interaction between two species. Biometrika **47**, 219–234 (1960)
14. Sokol, W., Howell, J.A.: The kinetics of phenol oxidation by washed cells. Biot. Bioe. **30**, 921–927 (1987)
15. Andrews, J.F.: A mathematical model for continuous culture of microorganisms utilizing inhibitory substrates. Biotechnol. Bioeng. **10**, 707–723 (1968)
16. Collings, J.B.: The effect of the functional response on the bifurcation behavior of a mite predator-prey interaction model. J. Math. Biol. **36**, 149–168 (1997)
17. Holling, C.S.: The functional response of predators to prey density and its rule in mimicry and population regulation. Mem. Entomology. Soc. Can. **45**, 3–60 (1965)
18. Ruan, S., Xiao, D.: Global analysis in a predator-prey system with non-monotonic functional response. SIAM J. Appl. Math. **61**(4), 1445–1472 (2001)
19. Zengyun, Hu, Teng, Zhidong, Zhang, Long: Stability and bifurcation analysis of a discrete predator-prey model with nonmonotonic functional response. Nonlinear Anal. Real World Appl. **12**, 2356–2377 (2011)
20. Priyadarshi, A., Gakkhar, S.: Dynamics of Leslie–Gower type generalist predator in tri-trophic food web system. Commun. Nonlinear Sci. Simulat. **18**, 3202–3218 (2013)
21. Letellier, C., Aguirre, L., Maquet, J., Aziz-Alaoui, M.A.: Should all the species of a food chain be counted to investigate the global dynamics. Chaos Solitons Fractals **13**(5), 1099–1113 (2002)
22. Upadhyay, R.K., Raw, S.N.: Complex dynamics of a three species food chain model with Holling Type IV functional response. Nonlinear Anal. Model. Control **16**(3), 353–374 (2011)
23. Shen, C.: Permanence and global attractivity of the food chain system with Holling type IV Functional response. Appl. Math. Comput. **215**, 179–185 (2009)
24. Hall, J.K.: Ordinary Differential Equation. Wiley-Interscience, New-York (1969)
25. May, R.M.: Stability and Complexity in Model Ecosystems. Princeton University Press, Princeton (1973)
26. Gerald, C.F., Wheatly, P.O.: Applied Numerical Analysis. Adison-Wesley, New York (2004)

Fifth-Order Four-Stage Explicit Trigonometrically-Fitted Runge–Kutta–Nyström Methods

Musa A. Demba, Norazak Senu and Fudziah Ismail

Abstract In this study, we derive fifth-order four-stage explicit trigonometrically-fitted Runge–Kutta–Nyström (ETFRKN) methods for the numerical integration of second-order initial value problems with oscillatory solutions based on Simos technique. The numerical results show the efficiency of the proposed methods in comparison with other Runge–Kutta–Nyström (RKN) Methods.

Keywords Trigonometric fitting · RKN methods · Oscillatory solutions · Numerical integration · Initial value problems

1 Introduction

During the last decades, methods for the numerical integration of initial value problems

$$y'' = f(x, y), \qquad x \in [x_0, X], \tag{1}$$

$$y(x_0) = y_0, \qquad y'(x_0) = y'_0,$$

whose solution shows a noticeable oscillatory behavior has attracted the interest of many researchers. Such problems occur in several fields of applied sciences; for instance, molecular dynamics, celestial mechanics, theoretical physics, and electronics. Bettis [1] suggested the first Runge–Kutta (RK) methods with 3 and 4 stages for the solution of ordinary differential equations (ODEs) with oscillatory solutions. Recently, Monovasilis [2], J.M. Franco [3], R.D'Ambrosio [4], H. Ramos [5]

M.A. Demba (✉) · N. Senu · F. Ismail
Department of Mathematics, Institute for Mathematical Research,
Universiti Putra Malaysia (UPM), 43400 Serdang, Selangor, Malaysia
e-mail: musdem2004@gmail.com

N. Senu
e-mail: norazak@upm.edu.my

F. Ismail
e-mail: fudziah@upm.edu.my

A. Kılıçman et al. (eds.), *Recent Advances in Mathematical Sciences*,
DOI 10.1007/978-981-10-0519-0_3

proposed Runge–Kutta–Nyström (RKN) methods for the solution of second-order ordinary differential equations. Similarly, T.E. Simos [6], J.P. Coleman [7] constructed an explicit RK method which integrate certain first-order initial value problems with periodic or oscillating solutions. In the same manner, G.V. Berghe [8] proposed exponentially-fitted RK (EFRK) methods that integrates exactly first order systems whose solutions can be represented as the linear combination of some functions. Motivated by the work of Simos [9], we construct fifth-order four-stage explicit trigonometrically-fitted RKN method. The remaining part of this paper is structured as follows: Sect. 2 deals with the derivation of the proposed method.

In Sect. 3, we present the numerical results and the last section deals with the conclusion.

2 Derivation of the Proposed Method

In this section, we will derive four-stage fifth-order explicit trigonometrically-fitted RKN methods using Simos technique. Let us consider the four-stage explicit RKN method given by:

$$y_{n+1} = y_n + hy'_n + h^2 \sum_{i=1}^{4} b_i f(x_n + c_i h, Y_i), \tag{2}$$

$$y'_{n+1} = y'_n + h \sum_{i=1}^{4} d_i f(x_n + cih, Y_i), \tag{3}$$

$$Y_i = y_n + c_i hy'_n + h^2 \sum_{j=1}^{3} a_{ij} f(x_n + cih, Y_i). \tag{4}$$

or in Butcher Tableau as :

0				
c_2	a_{21}			
c_3	a_{31}	a_{32}		
c_4	a_{41}	a_{42}	a_{43}	
	b_1	b_2	b_3	b_4
	d_1	d_2	d_3	d_4

In this study, the four-stage fifth-order dispersive of order eight RKN methods will be used as given in [10]. The coefficients of the methods are given in Tables 1 and 2.

When an explicit Runge–Kutta–Nyström method (2)–(4) is applied to the ODE (1), the method becomes:

Table 1 The RKN4(5,8,5)M Method [10]

0				
$\frac{1}{2}$	$\frac{1}{8}$			
$\frac{19}{70}$	$\frac{2907}{343000}$	$\frac{1216}{42875}$		
$\frac{44}{51}$	$\frac{6624772}{12858819}$	$\frac{6273905}{54121608}$	$\frac{210498365}{1028310552}$	
	$\frac{479}{5016}$	$\frac{235}{1776}$	$\frac{145775}{641744}$	$\frac{309519}{6873416}$
	$\frac{476}{5016}$	$\frac{235}{888}$	$\frac{300125}{962616}$	$\frac{2255067}{6873416}$

Table 2 The RKN4(5,8,5)S Method [10]

0				
$\frac{34}{105}$	$\frac{578}{11025}$			
$\frac{121}{142}$	$\frac{10640377}{194703584}$	$\frac{60046371}{194703584}$		
$\frac{1}{2}$	$\frac{1576823}{23630816}$	$\frac{5787873}{1538346592}$	$\frac{18651700}{342169003}$	
	$\frac{2623}{24684}$	$\frac{5479425}{19818532}$	$\frac{2505377}{47655850}$	$\frac{359}{5550}$
	$\frac{2623}{24684}$	$\frac{8103375}{19818532}$	$\frac{25411681}{71483775}$	$\frac{359}{2775}$

$$y_{n+1} = y_n + hy'_n + h^2 \sum_{i=1}^{4} b_i f(x_n + c_i h, Y_i), \tag{5}$$

$$y'_{n+1} = y'_n + h \sum_{i=1}^{4} d_i f(x_n + cih, Y_i), \tag{6}$$

with

$$Y_1 = y_n + c_1 hy'_n, \tag{7}$$

$$Y_2 = y_n + c_2 hy'_n - h^2 a_{21} w^2 Y_1, \tag{8}$$

$$Y_3 = y_n + c_3 hy'_n + h^2(-a_{31} w^2 Y_1 - a_{32} w^2 Y_2), \tag{9}$$

$$Y_4 = y_n + c_4 hy'_n + h^2(-a_{41} w^2 Y_1 - a_{42} w^2 Y_2 - a_{43} w^2 Y_3), \tag{10}$$

which results in

$$y_{n+1} = y_n + hy'_n + h^2 \sum_{i=1}^{4} b_i(-w^2 Y_i), \tag{11}$$

and

$$y'_{n+1} = y'_n + h \sum_{i=1}^{4} d_i(-w^2 Y_i). \tag{12}$$

Now, let $y_n = e^{Iwx}$. Computing the value of y_{n+1}, y'_n, and y'_{n+1} and substituting in the Eqs. (7)–(12) and by using $e^{Iv} = \cos(v) + I\sin(v)$ and comparing the real and imaginary part, we obtain the following system of equations:

$$T_1 := \cos(v) = 1 - v^2 \sum_{i=1}^{4} b_i \left(1 - v^2 \sum_{j=1}^{3} a_{ij} Y_j e^{-Iwx}\right), \tag{13}$$

$$T_2 := \sin(v) = v - v^2 \sum_{i=1}^{4} b_i c_i v, \tag{14}$$

$$T_3 := \sin(v) = v \sum_{i=1}^{4} d_i \left(1 - v^2 \sum_{j=1}^{3} a_{ij} Y_j e^{-Iwx}\right), \tag{15}$$

$$T_4 := \cos(v) = 1 - v^2 \sum_{i=1}^{4} d_i c_i. \tag{16}$$

where $v = wh$.

Solving (13)–(16) using the coefficients of the method in Table 1 for a_{31}, c_2, b_2, d_2, we obtain the Taylor series expansion of the solution as given in (17).

$$\begin{aligned}
a_{31} &= \frac{2907}{343000} + \frac{922507}{612255000}v^2 + \frac{1232013673651}{61638159870000000}v^4 \\
&\quad - \frac{17236094463540793}{1264054836560670000000}v^6 - \frac{1038588659926458350176663}{8271734680034077952700000000000}v^8 \\
&\quad + \frac{4494075784581591760180534261}{30281767897372755047640720000000000}v^{10} + O\left(v^{12}\right), \\
d_2 &= \frac{235}{888} + \frac{875000}{2577531}v^2 + \frac{23}{48960}v^4 + \frac{41078983}{690059865600}v^6 \\
&\quad - \frac{39685575187087}{530681913891360000}v^8 - \frac{4033241180098426998 7}{926048430651045461760000000}v^{10} \\
&\quad + \frac{177870063525579843149008 01}{2330724992684083770430656000000000}v^{12} + O\left(v^{14}\right), \\
c_2 &= \frac{1}{2} + \frac{25567}{33558000}v^4 - \frac{2406857437}{33784180920000}v^6 - \frac{118742174747143}{277133888365488000000}v^8 \\
&\quad + \frac{380049278740603078721}{7254046040099856117120000000}v^{10} + \frac{12003758502069503990362649}{486862554027341943156625920000000}v^{12} \\
&\quad - \frac{108961553964388992075619509340 3}{32676267176099081856900105246720000000000}v^{14} + O\left(v^{15}\right),
\end{aligned}$$

$$
\begin{aligned}
b_2 = {} & \frac{235}{1776} + \frac{6125000}{131454081}v^2 - \frac{227}{3427200}v^4 + \frac{89457317}{287524940000}v^6 \\
& - \frac{98730120494347}{1273636593339264000000}v^8 - \frac{1595918038577012384 21}{6173656204340303078400000000}v^{10} \\
& + \frac{645599919558530114442893}{4972213317726045376918732800000 0}v^{12} + O\left(v^{13}\right).
\end{aligned} \tag{17}
$$

This result in the new method called ETFRKN5(4)M. In a similar way, solving the above system using the coefficients of the method in Table 2 for a_{21}, a_{31}, c_2, c_3, we obtain the Taylor series expansion of the solution as given in (18).

$$
\begin{aligned}
a_{21} = {} & \frac{578}{11025} + \frac{848521}{243101250}v^2 - \frac{126992892839}{507449549250000}v^4 + \frac{986801812486171}{1765416981840750000000}v^6 \\
& - \frac{372016222314912877 21}{5527697111841572325000000000}v^8 + \frac{4263632051153968038490150 1}{825003819014954012091157500000000000}v^{10} \\
& - \frac{1093403999469958113112606 7711}{391389311775412501099791401250000000000000}v^{12} + O\left(v^{14}\right), \\
a_{31} = {} & \frac{10640377}{194703584} - \frac{11495}{2863288}v^2 + \frac{608371181}{398455158080}v^4 - \frac{154892867010409}{1871404418196432000}v^6 \\
& + \frac{1151397714695854901}{651061597090538692800000}v^8 - \frac{515488060970115012286 1}{2453796904301149454938800000000}v^{10} \\
& + \frac{2958528130879689827642928239}{1843940036893758974006130403200000000 0}v^{12} + O\left(v^{14}\right), \\
c_2 = {} & \frac{34}{105} - \frac{1629739}{1380815100}v^4 + \frac{354787385213}{864694031922 0000}v^6 - \frac{64967657425690231}{992729277228690540000 00}v^8 \\
& + \frac{83374602989572618 84051}{134694501063665961157740000000 0}v^{10} - \frac{5561821852109702943117 1117}{1405806507601481636603332380000000000 0}v^{12} \\
& + \frac{3357253776185608367982043 9733}{1834050314979582980153622506257500000000000}v^{14} + O\left(v^{15}\right), \\
c_3 = {} & \frac{121}{142} + \frac{1001275}{304940172}v^4 - \frac{10628432485}{25461284601312}v^6 + \frac{23345592719251}{1771596182559288960}v^8 \\
& - \frac{148060038779758582 73}{7211158249374806561452800 0}v^{10} + \frac{3716634862365696662369}{1929816888428842463638022400000}v^{12} \\
& - \frac{454933058077823261288939059}{3709392707551278469375665661104 0000000}v^{14} + O\left(v^{15}\right).
\end{aligned} \tag{18}
$$

This result in the new method called ETFRKN5(4)S.

3 Numerical Results

We compare the performance of the proposed methods ETFRKN5(4)M and ETFRKN5(4)S with RKN 4(5,8,5)M and RKN 4(5,8,5)S by considering the following problems:

Problem 1 [11]

$$
y'' = -64y, \qquad x \in [0, 4000], \tag{19}
$$

$$
y(0) = 1, \qquad y'(0) = -2,
$$

The exact solution is $y(x) = -\frac{1}{4}\sin(8x) + \cos(8x)$.
Problem 2 [12]

$$y'' = -100y, \qquad x \in [0, 4000], \tag{20}$$

$$y(0) = 1, \qquad y'(0) = 2,$$

The exact solution is $y(x) = -\frac{1}{5}\sin(10x) + \cos(10x)$.
Problem 3 [13]

$$y'' = -100y + 99\sin(x), \qquad x \in [0, 4000], \tag{21}$$

Table 3 Numerical results for problem 1

h	Methods	T = 100	T = 1000	T = 4000
0.025	RKN4(5,8,5)M	3.692168(−6)	3.679147(−5)	1.472381(−4)
	ETFRKN 5(4)M	4.561014(−11)	5.243194(−9)	8.3778224(−8)
	RKN 4(5,8,5)S	9.195562(−6)	9.268534(−5)	3.713213(−4)
	ETFRKN 5(4)S	4.702866(−11)	5.228975(−9)	8.383908(−8)
0.05	RKN4(5,8,5)M	1.167864(−4)	1.162936(−3)	4.663103(−3)
	ETFRKN 5(4)M	6.756228(−10)	7.937932(−9)	4.879474(−8)
	RKN 4(5,8,5)S	3.048566(−4)	3.100016(−3)	1.236002(−2)
	ETFRKN 5(4)S	1.178220(−9)	9.256040(−9)	9.638870(−8)
0.1	RKN4(5,8,5)M	3.591201(−3)	3.595222(−2)	1.515525(−1)
	ETFRKN 5(4)M	1.647945(−6)	1.644000(−5)	6.588576(−5)
	RKN 4(5,8,5)S	1.152519(−2)	1.108431(−1)	3.768479(−1)
	ETFRKN 5(4)S	1.331787(−6)	1.371467(−5)	5.485441(−5)

Table 4 Numerical results for problem 2

h	Methods	T = 100	T = 1000	T = 4000
0.025	RKN4(5,8,5)M	1.391246(−5)	1.386496(−4)	5.54318(−4)
	ETFRKN 5(4)M	5.377163(−11)	6.503970(−9)	1.035798(−7)
	RKN 4(5,8,5)S	3.504919(−5)	3.533674(−4)	1.412718(−3)
	ETFRKN 5(4)S	6.990865(−11)	6.341439(−9)	1.042249(−7)
0.05	RKN4(5,8,5)M	4.346169(−4)	4.358384(−3)	1.753295(−2)
	ETFRKN 5(4)M	9.798471(−9)	9.865613(−8)	3.914656(−7)
	RKN 4(5,8,5)S	1.196191(−3)	1.205331(−2)	4.736293(−2)
	ETFRKN 5(4)S	1.431150(−8)	1.395804(−7)	6.318411(−7)
0.1	RKN4(5,8,5)M	1.2885867(−2)	1.368225(−1)	6.672607(−1)
	ETFRKN 5(4)M	2.658254(−5)	2.701145(−4)	1.080911(−3)
	RKN 4(5,8,5)S	4.821748(−2)	3.924927(−1)	8.755385(−1)
	ETFRKN 5(4)S	1.653645(−5)	1.675094(−4)	6.785474(−4)

$$y(0) = 1, \qquad y'(0) = 11,$$

The exact solution is $y(x) = \sin(10x) + \cos(10x) + \sin(x)$.
Problem 4 [11]

$$y'' = -y + x, \qquad x \in [0, 4000], \tag{22}$$

$$y(0) = 1, \qquad y'(0) = 2,$$

The exact solution is $y(x) = \cos(x) + \sin(x) + x$.

The numerical results are presented in Tables 3, 4, 5 and 6.
The performance of these methods are presented graphically in Figs. 1, 2, 3 and 4.

Table 5 Numerical results for problem 3

h	Methods	T = 100	T = 1000	T = 4000
0.025	RKN4(5,8,5)M	1.891016(−5)	1.915843(−4)	7.684370(−4)
	ETFRKN 5(4)M	4.817489(−10)	9.075352(−9)	1.439351(−7)
	RKN 4(5,8,5)S	4.940767(−5)	4.899236(−4)	1.960017(−3)
	ETFRKN 5(4)S	1.927360(−9)	9.319327(−9)	1.456267(−7)
0.05	RKN4(5,8,5)M	5.902084(−4)	6.028942(−3)	2.429924(−2)
	ETFRKN 5(4)M	2.547192(−8)	1.476824(−7)	5.542078(−7)
	RKN 4(5,8,5)S	1.685789(−3)	1.674061(−2)	6.570742(−2)
	ETFRKN 5(4)S	9.968822(−8)	2.702436(−7)	9.540691(−7)
0.1	RKN4(5,8,5)M	1.737163(−2)	1.891174(−1)	9.247597(−1)
	ETFRKN 5(4)M	3.730664(−5)	3.752516(−4)	1.4977996(−3)
	RKN 4(5,8,5)S	6.763346(−2)	5.448985(−1)	1.214736(+0)
	ETFRKN 5(4)S	2.668692(−5)	2.345496(−4)	9.322934(−4)

Table 6 Numerical results for problem 4

h	Methods	T = 100	T = 1000	T = 4000
0.025	RKN4(5,8,5)M	1.665512(−11)	9.304131(−10)	1.431408(−8)
	ETFRKN 5(4)M	7.489120(−12)	9.009682(−10)	1.434046(−8)
	RKN 4(5,8,5)S	5.249490(−11)	1.032845(−9)	1.455373(−8)
	ETFRKN 5(4)S	8.071765(−12)	8.984671(−10)	1.45007(−8)
0.05	RKN4(5,8,5)M	5.229737(−10)	6.126811(−9)	2.593288(−8)
	ETFRKN 5(4)M	4.860112(−12)	5.171614(−10)	8.164534(−9)
	RKN 4(5,8,5)S	1.674536(−9)	1.551314(−8)	6.202617(−8)
	ETFRKN 5(4)S	4.718004(−12)	5.123866(−10)	8.172265(−9)
0.1	RKN4(5,8,5)M	1.670104(−8)	1.948985(−7)	7.882986(−7)
	ETFRKN 5(4)M	2.131628(−12)	2.229399(−10)	3.582954(−9)
	RKN 4(5,8,5)S	5.376775(−8)	4.969693(−7)	1.971801(−6)
	ETFRKN 5(4)S	1.662670(−12)	2.223715(−10)	3.512014(−9)

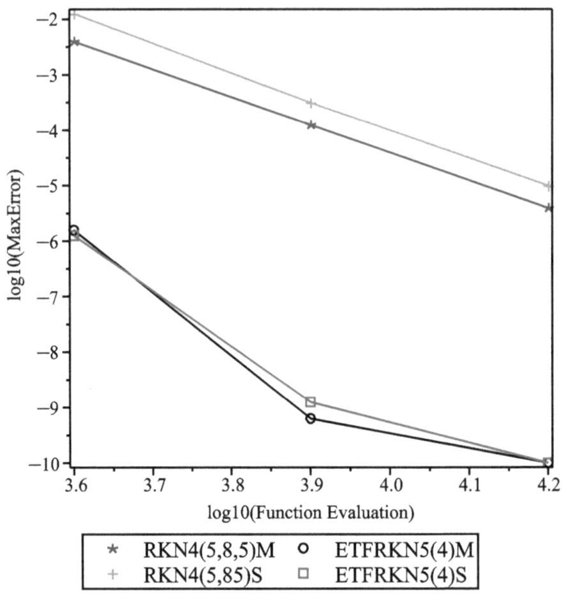

Fig. 1 Efficiency graph for problem 1 with $t_{end} = 100$ and $h = i(0.025)$, $i = 1, 2, 4$

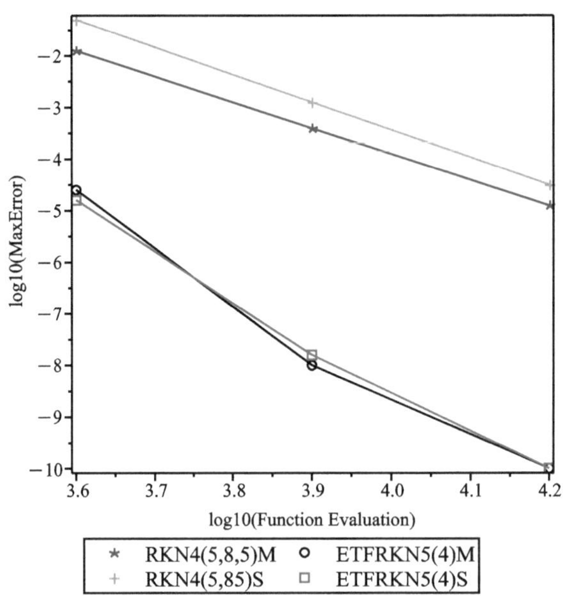

Fig. 2 Efficiency graph for problem 2 with $t_{end} = 100$ and $h = i(0.025)$, $i = 1, 2, 4$

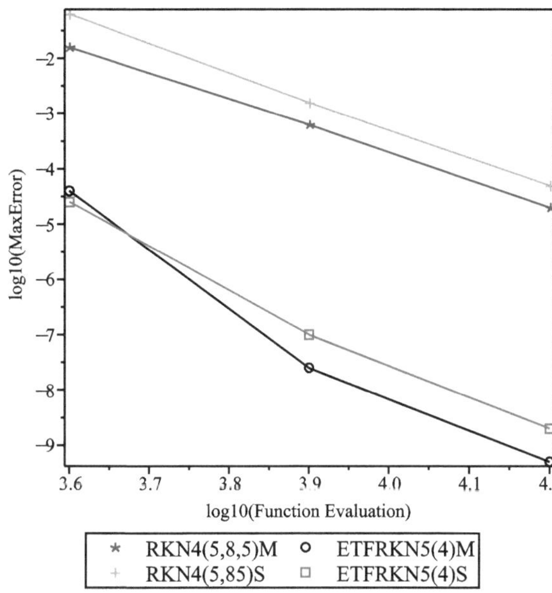

Fig. 3 Efficiency graph for problem 3 with $t_{end} = 100$ and $h = i(0.025)$, $i = 1, 2, 4$

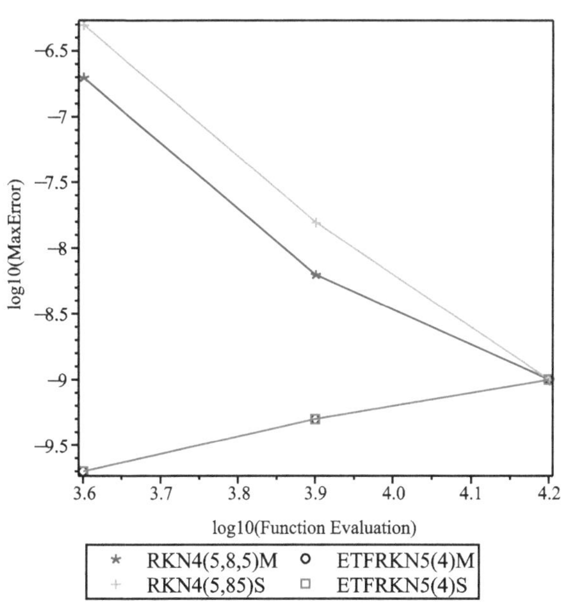

Fig. 4 Efficiency graph for problem 4 with $t_{end} = 1000$ and $h = i(0.025)$, $i = 1, 2, 4$

4 Conclusion

In this study, we have presented the fifth-order four-stage explicit trigonometrically-fitted RKN methods for the solutions of oscillatory problems. The numerical results obtained show that the error norm of the new methods are smaller than that of the other existing methods; the new methods are more efficient than the other existing methods.

Acknowledgments We are thankful to the Institute of Mathematical Research (INSPEM) and the Department of Mathematics, Universiti Putra Malaysia for the support and assistance during the research work.

References

1. Bettis, D.: Zeitschrift für angewandte Mathematik und Physik ZAMP **30**(4), 699 (1979)
2. Monovasilis, T., Kalogiratou, Z., Simos, T.: Appl. Math. Inf. Sci. **7**(1), 81 (2013)
3. Franco, J., Gómez, I.: J. Comput. Appl. Math. **260**, 482 (2014)
4. D'Ambrosio, R., Paternoster, B., Santomauro, G.: Appl. Math. Lett. **30**, 56 (2014)
5. Ramos, H., Vigo-Aguiar, J.: Comput. Phys. Commun. **185**(4), 1230 (2014)
6. Simos, T.: Comput. Phys. Commun. **115**(1), 1 (1998)
7. Coleman, J.P., Duxbury, S.C.: J. Comput. Appl. Math. **126**(1), 47 (2000)
8. Berghe, G.V., De Meyer, H., Van Daele, M., Van Hecke, T.: Comput. Phys. Commun. **123**(1), 7 (1999)
9. Simos, T.E.: Appl. Math. Lett. **15**(2), 217 (2002)
10. Senu, N.: Runge–Kutta–Nystrom methods for the solutions of oscillatory differential equations. Ph.D. Dissertation, Department of mathematics, Faculty of Science, 43400 Serdang, Malaysia (2009)
11. Ahmad, S., Ismail, F., Senu, N., Suleiman, M.: Appl. Math. Comput. **219**(19), 10096 (2013)
12. Senu, N., Suleiman, M., Ismail, F., Othman, M.: WSEAS Trans. Math. **9**(9), 679 (2010)
13. Anastassi, Z., Kosti, A.: J. Comput. Appl. Math. **275**, 311 (2015)

Modified Homotopy Perturbation Method for Fredholm–Volterra Integro-Differential Equation

F.S. Zulkarnain, Z.K. Eshkuvatov, N.M.A. Nik Long and F. Ismail

Abstract In this paper, consider a linear Fredholm–Volterro integro-differential equation (FVIDE) of the third kind has derivative of order m where m is positive integer. This type of integral has been solved by using modified homotopy perturbation method (HPM) to get approximate solutions. In this modification, selective functions and unknown parameters are introduced to help us obtain only two-step iterations. This proposed method could avoid common problems such as complex and long calculations. It is found that modified HPM is a semi-analytical method and easy to apply for solving FVIDE. Numerical examples are given to present the efficiency and reliability of the propose method.

Keywords Homotopy perturbation method · Numerical method · Integral equation · Approximate solution · Integro-differential equations

1 Introduction

Homotopy perturbation method (HPM) [1–5] is the combination of two methods namely homotopy and perturbation methods. Recently, researchers have been using HPM to solve mathematic and physic problems. This method deforms a complicated problem to a simple problem which is easy to solve.

HPM has been used for finding the exact and approximate solutions; in general, mathematical problems such as nonlinear ordinary differential equations (ODEs) [6],

F.S. Zulkarnain (✉) · F. Ismail
Faculty of Science, Departments of Mathematics, Universiti Putra Malaysia (UPM), Serdang, Selangor, Malaysia
e-mail: fsamihah88@gmail.com

Z.K. Eshkuvatov
Faculty of Science and Technology, Universiti Sains ISlam Malaysia (USIM), Nilai, Negeri Sembilan, Malaysia

N.M.A. Nik Long
Institute for Mathematical Research, Universiti Putra Malaysia (UPM), Serdang, Selangor, Malaysia

A. Kılıçman et al. (eds.), *Recent Advances in Mathematical Sciences*,
DOI 10.1007/978-981-10-0519-0_4

one-phase inverse Stefan problem [7], linear and nonlinear integral equations [8], the integro-differential equations [9, 10] and the Volterra–Fredholm integral equations [11].

There are few modifications on HPM, one of them is adding few unknown parameters to the first iterations and finding them by equating the second iteration to be zero that leads to semi-analytical solutions [12]. Another modification is to divide the interval into subintervals and use HPM in each subinterval which is named multistage homotopy-perturbation method [13, 14]. A series of parameters and selectives called improved HPM have been added to find semi-analytical solutions of nonlinear Fredholm and Volterra integral equations [15]. Mohamad Nor et al. [16] developed the new homotopy function using De Casteljau algorithms to solve algebraic nonlinear problems.

Our main problem is to implement modified HPM (MHPM) into Fredholm–Volterra integro-differential equation of the third kind. Fredholm–Volterra integral equations arise from parabolic boundary value problems, the mathematical modelling of spatiotemporal development of an epidemic and various physical and biological problems [17, 18]. There are many methods to solve the Volterra–Fredholm integral equations; for instance, the collocation method [19], the Galerkin Method [20], the fixed point method [21], the moving least squares method [22], the Taylor expansion method [23], and the modified decomposition method [24]. Consider Fredholm–Volterra integro-differential equation of the third kind with initial conditions.

$$s(x)\,u^{(m)}(x) = f(x) + \lambda_1 \int_a^b K_1(x,t)\,u(t)\,dt + \lambda_2 \int_a^x K_2(x,t)\,u(t)\,dt, \quad x \in [a,b],$$

$$u^{(k)}(a) = c_k, \; 0 \le k \le m-1. \quad (1)$$

where $s(x)$, $g(x)$ are continuous functions on $[a, b]$, m is the order of differential, λ_1 and λ_2 are parameters, K_1 and K_2 are square integrable kernels, and $u(x)$ is the function to be determined.

Rewrite Eq. (1) in thc operator form as

$$Lu = f + Ku, \quad (2)$$

where L is differential operator and K is integral operator has the form

$$Lu = u^{(m)}(x), \qquad f = \frac{f(x)}{s(x)},$$

$$Ku = \frac{\lambda_1}{s(x)} \int_a^b K_1(x,t)\,u(t)\,dt + \frac{\lambda_2}{s(x)} \int_a^x K_2(x,t)\,u(t)\,dt.$$

Assume that L^{-1} is integral operator and has the form

$$L^{-1}(\cdot) = \int_a^x \int_a^{x_1} \cdots \int_a^{x_{m-1}} (\cdot) dx_1 dx_2 \ldots dx_{m-1}, \quad (3)$$

Applying (3) to the Eq. (2) and taking into account initial condition in Eq. (1), we obtain

$$u(x) = \sum_{k=0}^{m-1} \frac{c_k}{k!}(x-a)^k + L^{-1}\left[\frac{f(x)}{s(x)}\right] + L^{-1}\left[\frac{\lambda_1}{s(x)}\int_a^b K_1(x,t)\,u(t)\,dt + \frac{\lambda_2}{s(x)}\int_a^x K_2(x,t)\,u(t)\,dt\right]. \quad (4)$$

Operator form of Eq. (4) is

$$L_m u = h_m + f_m + K_m u, \quad (5)$$

where

$$L_m u = u(x), \quad h_m = \sum_{k=0}^{m-1} \frac{c_k}{k!}(x-a)^k, \quad f_m = L^{-1}\left(\frac{f(x)}{s(x)}\right),$$

$$K_m u = L^{-1}\left(\frac{\lambda_1}{s(x)}\int_a^b K_1(x,t)\,u(t)\,dt + \frac{\lambda_2}{s(x)}\int_a^x K_2(x,t)\,u(t)\,dt\right).$$

2 Modified Homotopy Perturbation Method

Standard HPM is usually given by

$$H(v,p) = (1-p)F(v) + p(L_m v - f_m - K_m v), \quad (6)$$

where $F(v)$ is a functional operator with known solution u_0, which can be obtained easily.

The convex homotopy (6) continuously traces an implicitly defined curve from a starting point $H(u(x),0)$ to a solution function $H(u(x),1)$. The embedding parameter p monotonically increases from zero to unit as trivial problem $F(u)=0$ is continuously deformed to original problem $L_m u - f_m - K_m u = 0$. It is clear that,

$$H(v,p) = 0, \quad (7)$$

leads to

$$F(v) = p[F(v) + L_m v - f_m - K_m v]. \quad (8)$$

Improved HPM has been presented in Ghorbani [15]. Using this method, we define $F(v) = L_m v - \sum_{j=0}^{N} \alpha_j\, g_j(x)$ where α_j are called accelerator parameters, and using these parameters we might have exact solution or approximate solution. Thus, the improved HPM is constructed as follows:

$$H_m(v,\alpha,p) = (1-p)\left(L_m v - \sum_{j=0}^{N} \alpha_j\, g_j(x)\right) + p\,(L_m v - h_m - f_m - K_m v), \tag{9}$$

where α_j are the parameters to be defined and $g_j(x)$ are given selective functions. Forcing $H_m(v,\alpha,p) = 0$ leads to the equation

$$L_m v = \sum_{j=0}^{N} \alpha_j\, g_j(x) + p\left[h_m + f_m + K_m v - \sum_{j=0}^{N} \alpha_j\, g_j(x)\right]. \tag{10}$$

Let us search approximate solution v in the form of a power series

$$v(x) = \sum_{k=0}^{\infty} p^k v_k(x). \tag{11}$$

Substituting (11) into (10) yields

$$L_m\left[\sum_{k=0}^{\infty} p^k v_k(x)\right] = \sum_{j=0}^{N} \alpha_j\, g_j(x) + p\left(h_m + f_m + K_m\left[\sum_{k=0}^{\infty} p^k v_k(x)\right] - \sum_{j=0}^{N} \alpha_j\, g_j(x)\right). \tag{12}$$

Since $L_m = I$ is identity operator, comparing the expressions from both sides of Eq. (12) like power of parameter p, we get the following schemes

$$\begin{aligned} v_0 &= \sum_{j=0}^{N} \alpha_j\, g_j(x), \\ v_1 &= h_m + f_m + K_m v_0 - \sum_{j=0}^{N} \alpha_j\, g_j(x), \\ v_k &= K_m v_{k-1},\ k \geq 2. \end{aligned} \tag{13}$$

Remark: In MHPM, the accelerating parameters α_j are defined by forcing $v_1 = 0$ it leads two-step iteration and gives exact solution. If $v_1 \neq 0$ but $v_1^N \to 0$ as $N \to \infty$ then the contribution of $v_k, k \geq 2$ to the solution will be small; therefore, we can neglect $v_k, k \geq 2$ to find the approximate solution.

3 Results

Example 1 Consider 2nd order FVIDE as

$$(x+1)u''(x) = -\frac{119}{60} - \frac{71}{36}x + \frac{1}{2}x^2 - \frac{2}{3}x^3 + \frac{5}{12}x^4 - \frac{1}{10}x^5 + \frac{1}{3}\int_1^2 xt_0 u(t)\,dt$$
$$+\frac{1}{2}\int_1^x (x-t^2)u(t)\,dt, \quad u(1)=0,\ u'(1)=0 \tag{14}$$

which has exact solution $u(x) = -x^2 + 2x - 1$.

After reducing Eq. (14) into Fredholm–Volterra integral equation, we have

$$u(x) = \frac{707}{3600} - \frac{301}{180}\ln(2) + \frac{301}{180}\ln(x+1)\frac{301}{180} + \ln(x+1)\,x + \frac{5137}{3600}x$$
$$-\frac{301}{180}\ln(2)\,x - \frac{329}{180}x^2 + \frac{101}{360}x^3 - \frac{71}{720}x^4 - \frac{31}{1200}x^5 - \frac{1}{300}x^6$$
$$+\frac{1}{3}\int_1^x\int_1^{t_0}\int_1^2 \frac{t_1 t_2}{t_1+1}u(t_2)\,dt_2\,dt_1\,dt_0 + \frac{1}{2}\int_1^x\int_1^{t_0}\int_1^{t_1}\frac{t_1-t_2^2}{t_1+1}u(t_2)\,dt_2\,dt_1\,dt_0.$$

By using MHPM scheme (13), we choose selective function as $g_j(x) = x^j$ with $N = 4$. To find $\alpha = (\alpha_0, \alpha_1, \ldots, \alpha_4)$, we force $v_1 = 0$ and obtain system algebraic as below

$$-\frac{1}{784}\alpha_4 = 0,$$
$$\frac{1}{245}\alpha_4 - \frac{1}{504}\alpha_3 = 0,$$
$$-\frac{1}{175}\alpha_4 + \frac{1}{144}\alpha_3 - \frac{1}{300}\alpha_2 - \frac{1}{300} = 0,$$
$$\frac{3}{350}\alpha_4 - \frac{1}{96}\alpha_3 + \frac{1}{75}\alpha_2 - \frac{1}{160}\alpha_1 + \frac{31}{1200} = 0,$$
$$-\frac{71}{70}\alpha_4 + \frac{5}{288}\alpha_3 - \frac{1}{72}\alpha_2 + \frac{1}{32}\alpha_1 - \frac{1}{72}\alpha_0 - \frac{71}{720} = 0,$$
$$\frac{1}{35}\alpha_4 - \frac{149}{144}\alpha_3 + \frac{2}{45}\alpha_2 - \frac{1}{16}\alpha_1 + \frac{1}{9}\alpha_0 - \frac{101}{360} = 0,$$
$$\frac{113}{70}\alpha_4 + \frac{43}{40}\alpha_3 - \frac{71}{120}\alpha_2 + \frac{65}{144}\alpha_1 - \frac{1}{3}\alpha_0 - \frac{329}{180} = 0,$$
$$-\frac{221}{70}\alpha_4 - \frac{31}{15}\alpha_3 - \frac{43}{60}\alpha_2 - \frac{7}{9}\alpha_1 + \frac{5}{6}\alpha_0 + \frac{301}{180} = 0,$$
$$-\frac{221}{70}\alpha_4 - \frac{31}{15}\alpha_3 - \frac{43}{60}\alpha_2 - \frac{7}{9}\alpha_1 + \frac{5}{6}\alpha_0 + \frac{301}{180} = 0,$$

$$\left(\frac{221}{70}\ln 2-\frac{311}{2450}\right)\alpha_4+\left(\frac{31}{15}\ln 2-\frac{7}{288}\right)\alpha_3+\left(\frac{43}{60}\ln 2-\frac{43}{225}\right)\alpha_2$$
$$+\left(\frac{7}{9}\ln 2-\frac{33}{32}\right)\alpha_1-\left(\frac{5}{6}\ln 2+\frac{4}{9}\right)\alpha_0-\frac{301}{180}\ln 2+\frac{5137}{3600}=0,$$
$$\left(\frac{221}{70}\ln 2-\frac{29543}{19600}\right)\alpha_4+\left(\frac{31}{15}\ln 2-\frac{10361}{10080}\right)\alpha_3+\left(\frac{43}{60}\ln 2-\frac{449}{1800}\right)\alpha_2$$
$$+\left(\frac{7}{9}\ln 2-\frac{551}{1440}\right)\alpha_1-\left(\frac{5}{6}\ln 2-\frac{23}{72}\right)\alpha_0-\frac{301}{180}\ln 2+\frac{707}{3600}=0. \quad (15)$$

We found that $\alpha_0=-1, \alpha_1=2, \alpha_2=-1$ and $\alpha_3=\alpha_4=0$. Thus, the approximate solution is $v(x)=v_0(x)=-x^2+2x-1$ is identical to the exact solution.

Example 2 Consider fourth-order FVIDE as

$$(e^x)\,u^{(4)}(x)=\frac{3}{40}e^x+\frac{1}{8}e^{x+1}-\frac{1}{2}e^{x-1}-\frac{2}{15}e^x x^3-\frac{1}{10}e^{3x}x+\frac{321}{20}e^{3x}$$
$$-\frac{1}{8}\int_0^1 e^{x-t}u(t)\,dt+\frac{1}{5}\int_0^x t\,e^x u(t)\,dt,$$
$$u(0)=1, u'(0)=4, u''(0)=4, u'''(0)=8. \quad (16)$$

with exact solution $u(x)=e^{2x}+2x$.

We choose functions $g_j(x)=x^j$ where $j=0,1,\ldots,N$ yields $v_0=\alpha_0+\alpha_1 x+\alpha_2 x^2+\cdots+\alpha_N x^N$. As we repeated the same calculations in Example 1, we cannot solve the algebraic equations for any N. Then, we substitute points $x_i, i=0,\ldots,N$ in the form

$$x_i=\frac{i\,(b-a)}{N}, \quad i=0,1,\ldots,N \quad (17)$$

Table 1 presented errors of $v(x)$ using MHPM for $N=\{5, 10, 15, 20\}$ in problem (16).

Example 3 Consider third-order FVIDE as

$$\frac{4}{2+\cos(2x)}u^{(3)}(x)=f(x)+\frac{3}{5}\int_0^{\pi}(x+t\sin(x)+\cos(t))\,u(t)\,dt$$
$$+\frac{1}{5}\int_0^x t\sin(x-t)\,u(t)\,dt,$$
$$u(0)=0, u'(0)=-\frac{1}{4}, \quad u''(0)=0, \quad (18)$$

$$f(x)=\frac{1}{9}+\frac{3}{5}\pi^2-\frac{1}{20}\pi^4 x+\frac{4}{5}x^2-\frac{1}{15}x^4-\left(\frac{3}{80}\pi+\frac{1}{25}\pi^5\right)\sin(x)-\frac{1}{60}x\sin(x)\cos(x)$$
$$+\frac{29}{18}\cos(x)-\frac{1}{45}\cos^2(x).$$

Table 1 Errors of modified HPM for Eq. (16)

N	$\|u(x) - v(x)\|$
5	3.0476237×10^{-4}
10	$6.3915328 \times 10^{-10}$
15	$1.8708452 \times 10^{-16}$
20	$1.3107657 \times 10^{-23}$

Table 2 Errors of modified HPM for Eq. (18)

N	$\|u(x) - v(x)\|$
5	1.5214140×10^{-3}
10	3.3985617×10^{-5}
15	$4.4100567 \times 10^{-10}$
20	$4.4531769 \times 10^{-14}$

Exact solution of Eq. (18) is $u(x) = \dfrac{\sin(-2x)}{8} + \dfrac{x^3}{3}$.

We choose selective function as $g_j(x) = x^j$ where $j = 0, 1, \ldots, N$ then applying the same calculations as Example 2. As a result, the errors are shown in Table 2. According to Tables 1 and 2, both the approximate solutions converge to the exact solution.

4 Conclusion

In this work, we are using MHPM to solve FVIDE at m order. The modified homotopy was introduced where $\alpha = [\alpha_j]$ is the unknown accelerating parameter and $g(x) = [g(x)_j]$ is the selective function. The unknown parameters are obtained by equating $v_1 = 0$ which lead $v_2 = v_3 = \cdots = v_k = 0$. In consequence, this method could avoid any long and complex computations as shown in the numerical examples. The numerical results present that the modified HPM is an efficient and reliable method for solving this type of integral equations.

Acknowledgments This work was supported by Universiti Putra Malaysia under Research Grant Universiti Putra Malaysia (Putra Grant 2014). The project code is GP-i/2014/9442300. Authors are grateful for the sponsor and financial support of the Research Management Center, Universiti Putra Malaysia.

References

1. He, J.-H.: Homotopy perturbation technique. Comput. Methods Appl. Mech. Eng. **178**, 257–262 (1999)
2. He, J.-H.: A coupling method of a homotopy technique and a perturbation technique for non-linear problems. Int. J. Nonlinear Mech. **35**, 37–43 (2000)

3. Khan, Y., Wu, Q.: Homotopy perturbation transform method for nonlinear equations using He's polynomials. Comput. Math. Appl. **61**, 1963–1967 (2011)
4. Madani, M., Fathizadeh, M., Khan, Y., Yildrim, A.: On coupling of the homotopy perturbation method and Laplace transformation. Math. Comput. Modell. **53**, 1937–1945 (2011)
5. Khan, Y., Akbarzade, M., Kargar, A.: Coupling of homotopy and the variational approach for a conservative oscillator with strong odd-nonlinearity. Sci. Iran. **19**, 417–422 (2012)
6. Ramos, J.I.: Piecewise homotopy methods for nonlinear ordinary differential equations. Appl. Math. Comput. **198**, 92–116 (2008)
7. Słota, D.: The application of the homotopy perturbation method to one-phase inverse Stefan problem. Int. Commun. Heat Mass Transf. **37**, 587–592 (2010)
8. Jafari, H., Alipour, M., Tajadodi, H.: Convergence of homotopy perturbation method for solving integral equations. Thai J. Math. **8**, 511–520 (2010)
9. Golbabai, A., Javidi, M.: Application of He's homotopy perturbation method for nth-order integro-differential equations. Appl. Math. Comput. **190**, 1409–1416 (2007)
10. Dehghan, M., Shakeri, F.: Solution of an integro-differential equation arising in oscillating magnetic fields using He's Homotopy Perturbation Method. Prog. Electromagn. Res. **78**, 361–376 (2008)
11. Ghasemi, M., Kajani, M.T., Davari, A.: Numerical solution of the nonlinear Volterra-Fredholm integral equations by using Homotopy Perturbation method. Appl. Math. Comput. **188**, 446–449 (2007)
12. Javidi, M., Golbabai, A.: Modified homotopy perturbation method for solving non-linear Fredholm integral equations. Chaos Solitons Fractals **40**, 1408–1412 (2009)
13. Chowdhury, M.S.H., Hashim, I.: Application of multistage homotopy-perturbation method for the solutions of the Chen system. Nonlinear Anal.: Real World. Appl. **10**(1), 381–391 (2009)
14. Chowdhury, M.S.H., Hassan, T.H., Mawa, S.: A new application of homotopy perturbation method to the reaction-diffusion Brusselator model. Proc.-Soc. Behav. Sci. **8**, 648–653 (2010)
15. Ghorbani, A., Saberi-Nadjafi, J.: Exact solutions for nonlinear integral equations by a modified homotopy perturbation method. Comput. Math. Appl. **28**, 1032–1039 (2008)
16. Mohamad Nor, H., Md Ismail, A.I., Abdul Majid, A.: A new homotopy function for solving nonlinear equations. AIP Conf. Proc. **1557**(21), 21–25 (2013)
17. Diekman, O.: Thresholds and traveling waves for the geographical spread of infection. J. Math. Biol. **6**, 109–130 (1978)
18. Thieme, H.R.: A model for the spatio spread of an epidemic. J. Math. Biol. **4**, 337 (1977)
19. Brunner, H.: On the numerical solution of nonlinear Volterra-Fredholm integral equation by collocation methods. SIAM J. Numer. Anal. **27**(4), 987–1000 (1990)
20. Hendi, F.A., Albugami, A.M.: Numerical solution for Fredholm-Volterra integral equation of the second kind by using collocation and Galerkin methods. J. King Saud Univ. (Sci.) **22**, 37–40 (2010)
21. Calió, F., Muñaz, M.V.F., Marchetti, E.: Direct and iterative methods for the numerical solution of mixed integral equations. Appl. Math. Comput. **216**, 3739–3746 (2010)
22. Dastjerdi, H.L., Ghaini, F.M.: Numerical solution of Volterra-Fredholm integral equations by moving least square method and Chebyshev polynomials. Appl. Math. Model. **36**, 3283–3288 (2012)
23. Chen, Z., Jiang, W.: An approximate solution for a mixed linear Volterra-Fredholm integral equation. Appl. Math. Lett. **25**, 1131–1134 (2012)
24. Bildik, N., Inc, M.: Modified decomposition method for nonlinear Volterra-Fredholm integral equations. Chaos Solitons Fractals **33**, 308–313 (2007)

Simultaneous Effects of Soret and Dufour on the Unsteady Stagnation Point Flow of Micropolar Fluid Towards a Permeable Stretching Sheet

Shah Jahan, Hamzah Sakidin and Roslinda Nazar

Abstract The present study investigates the Soret and Dufour effects on the unsteady stagnation point flow of micropolar fluid toward a permeable stretching surface. First, we obtained the ordinary differential equation using appropriate transformation on partial differential equations. Then we used the Homotopy analysis method (HAM) to solve the subsequent problem. The influences of different values for various pertinent parameters such as unsteady parameter, Schmidt number, chemical reaction parameter, Dufour and Soret number on concentration, and temperature distributions are determined. Numerical values of physically interested parameter like skin friction coefficient are computed and compared with the previous results.

Keywords Soret and dufour effect · Unsteady flow · Permeable stretching surface · Stagnation point flow · Micropolar fluid

1 Introduction

Theory which introduces the coupling between the spin of each particle and microscopic velocity is known as "the theory of micropolar fluids" introduced by Eringen [1]. This theory has a range of applications in colloidal suspensions, liquid crystals, turbulent, porous media, lubricants, blood flow, capillaries, and microchannels flows. Joneidi et al. [2] investigated the effect of high mass transfer on micropolar fluid flow in a porous channel. They solved it with optimal homotopy asymptotic

S. Jahan (✉) · H. Sakidin
Fundamental and Applied Sciences Department, Universiti Teknologi PETRONAS, 32610 Bandar Seri Iskander, Perak, Malaysia
e-mail: jahanshah669@gmail.com

H. Sakidin
e-mail: hamzah.sakidin@petronas.com.my

R. Nazar
Faculty of Science & Technology, School of Mathematical Sciences, Universiti Kebangsaan Malaysia, 43600 Bangi, Selongor, Malaysia
e-mail: rmn@ukm.edu.mu

A. Kılıçman et al. (eds.), *Recent Advances in Mathematical Sciences*,
DOI 10.1007/978-981-10-0519-0_5

method and got excellent comparison with numerical results. Ashraf et al. [3] studied the asymmetric laminar flow of micropolar fluid numerically in porous channel. They observed that micropolar fluids increase couple stress and reduce shear stress near the wall when compared with Newtonian fluids. Mapatra and Gupta examined the effect of heat transfer in stagnation point flow over a stretching sheet [4]. Then Nazar et al. [5] studied the micropolar fluid flow over a stretching sheet numerically. A phenomenon in which transfer of heat is induced by concentration gradient is called thermo-diffusion or Dufour effect. On the other hand the phenomenon of thermo-diffusion or Soret effect implies that mass transfer is induced by temperature gradient.

When concentration and temperature gradient are high, then these effects are very significant. The applications of these effects have become very prominent in the areas such as isotope separation, nuclear waste, hydrology, geothermal energy, and in the mixture of different gases like Helium or Hydrogen due to their light molecular weight. Similarly, for air or hydrogen due to medium molecular weight, Angel et al. [6] examined the effects of Dufour and Soret on free convection towards a vertical surface that was fixed in a porous medium. Cheng [7] studied the dual diffusion in a fluid-saturated porous medium adjacent to a perpendicular truncated cone with variable concentration and wall temperature considering the effect of Soret and Dufour. Mixed convection magnetohydrodynamic heat and mass transfer on a micropolar fluid towards a stretching surface saturated in porous medium under the impact of ohmic heating, Soret and Dufour have been discussed by Pal and Chatterjee [8]. Pal and Mondal [9–12] examined these effects in detail.

Motivated by the above studies, our aim is to examine simultaneous effects of Soret and Dufour on micropolar fluid past a permeable stretching sheet with unsteady stagnation point flow. Using suitable transformation, the nonlinear partial differential equations are transformed to ordinary differential equation. To solve highly nonlinear equations, we use Homotopy analysis method [13–15] to find the series solution. The influence of different flow parameters is discussed through graphs and compared numerically.

2 Problem Formulation

Consider the stagnation point flow of a micropolar fluid past a permeable stretching surface for unsteady case. The x-component of the surface is stretched with velocity U_w. As a result fluid moves with velocity U in the y-direction toward the stagnation point on the surface. The flow of heat and mass transfer with the effect of Soret and Dufour is also incorporated. With the help of boundary layer approximation, the associated equations for continuity, momentum, angular momentum, concentration, and energy are given as

$$\frac{\partial u}{\partial x} + \frac{\partial v}{\partial y} = 0, \tag{1}$$

$$\frac{\partial u}{\partial t} + u\frac{\partial u}{\partial x} + v\frac{\partial u}{\partial y} = U\frac{\partial U}{\partial x} + \frac{\partial U}{\partial t} + \left(\nu + \frac{\kappa}{\rho}\right)\frac{\partial^2 u}{\partial y^2} + \frac{\kappa}{\rho}\frac{\partial N^*}{\partial y}, \tag{2}$$

$$\frac{\partial N^*}{\partial t} + u\frac{\partial N^*}{\partial x} + v\frac{\partial N^*}{\partial y} = \frac{\gamma^*}{\rho j}\frac{\partial^2 N^*}{\partial y^2} - \frac{\kappa}{\rho j}\left(2N^* + \frac{\partial u}{\partial y}\right), \tag{3}$$

$$\frac{\partial C}{\partial t} + u\frac{\partial C}{\partial x} + v\frac{\partial C}{\partial y} = D_e\frac{\partial^2 C}{\partial y^2} - R(t)C + \frac{D_e k_T}{T_m}\frac{\partial^2 T}{\partial y^2}, \tag{4}$$

$$\frac{\partial T}{\partial t} + u\frac{\partial T}{\partial x} + v\frac{\partial T}{\partial y} = \alpha_m\frac{\partial^2 T}{\partial y^2} + \frac{D_e k_T}{C_s C_p}\frac{\partial^2 C}{\partial y^2}, \tag{5}$$

$$\left.\begin{array}{c} u = U_w,\ v = v_w,\ N^* = -N_0\frac{\partial u}{\partial y},\ T = T_w, \\ \mathrm{C} = \mathrm{C_w} \text{ at } y = 0, \\ u \to U,\quad N^* \to 0,\quad T \to T_\infty,\quad \mathrm{C} \to \mathrm{C}_\infty \text{ as } y \to \infty, \end{array}\right\} \tag{6}$$

where

$$\left.\begin{array}{c} U_w = \frac{ax}{1-ct},\quad U = \frac{bx}{1-ct},\quad v_w = -\sqrt{\frac{\nu U_w}{x}}f(0), \\ C_w = C_\infty + \frac{ex}{1-ct},\quad T_w = T_\infty + \frac{bx}{1-ct}. \end{array}\right\} \tag{7}$$

Note that $b > 0, a > 0, c \geq 0, e > 0,$ u and v are velocity components along x-axis and y-axis, $ct < 1, R(t) = R_1(1 - ct)^{-1}$ and R_1 is constant reaction rate, ρ is the fluid density, ν is the kinematic viscosity, C is the concentration, C_∞ is the free stream concentration,C_s is the concentration susceptibility, D_e is the coefficient of diffusing species, α_m is the thermal diffusivity, T is the temperature of the fluid, and T_∞ is the temperature for surrounding. U_w is the stretching velocity, v_w is the suction or injection velocity, C_w is the concentration at surface, T_w is the surface temperature, T_m is the mean fluid temperature, N^* is the microrotation or angular velocity, $j = (\nu/c)$ is the microinertia per unit mass, k is the thermal conductivity, k_T is the thermal diffusion rate, C_p is the specific heat, and $\gamma^* = (\mu + \kappa/2)j$ is the spin gradient viscosity, respectively.

Here $b > 0$ is for heated surface ($T_w > T_\infty$) and $b < 0$ for a cooled plate ($T_w < T_\infty$). Further, prime signifies the differentiation with respect to η. Obviously for $\gamma > 0$ or $\gamma < 0$, we have destructive/generative chemical reaction, whereas the case $\gamma = 0$ yields the nonreactive species. Note that for $k = 0$, we get viscous fluid. Further, the boundary parameter N_0 has range $0 \leq N_0 \leq 1$. For strong concentration ($N_0 = 0$) we get $N^* = 0$ near the wall. That mean due to saturated particle flows, microelements are unable to spin near the wall, whereas $N_0 = 1/2$ represents the weak concentration of microelements because stress tensor vanishes its anti-symmetric part. Peddieson [15] has shown that the turbulent boundary layer flow can be demonstrated when $N_0 = 1$, whereas $v_w < 0$ is for suction and $v_w > 0$ for injection/blowing.

Introducing

$$\left.\begin{array}{c}\eta = \left(\frac{U_w}{\nu x}\right)^{1/2} y, \quad u = cxf'(\eta), \quad \psi = -(U_w \nu x)^{1/2} f(\eta), \\ N^* = U_w \left(\frac{U_w}{\nu x}\right)^{1/2} g(\eta), \quad \phi(\eta) = \frac{C - C_\infty}{C_w - C_\infty}\end{array}\right\} \tag{8}$$

Now (1) is automatically satisfied and (2)–(6) are reduced as (9)–(13) respectively:

$$\left.(1+K)f''' + ff'' - (f')^2 + Kg' - \alpha\left(f' + \frac{1}{2}\eta f''\right) + \lambda^2 + \lambda\alpha = 0,\right\} \tag{9}$$

$$\left.\left(1 + \frac{K}{2}\right) g'' + fg' - f'g - 2Kg - Kf'' - \alpha\left(\frac{3}{2}g + \frac{1}{2}\eta g'\right) = 0,\right\} \tag{10}$$

$$\left.\phi'' - Sc\alpha(\phi + \frac{\eta}{2}\phi') + Sc(f\phi' - \phi f') - Sc\gamma\phi + ScSr\theta'' = 0,\right\} \tag{11}$$

$$\left.\theta'' - \Pr\alpha(\theta + \frac{\eta}{2}\theta') + \Pr(f\theta' - \theta f') + \Pr D_f\phi'' = 0,\right\} \tag{12}$$

$$\left.\begin{array}{c}f(0) = S, \quad f'(0) = 1, \quad g(0) = -N_0 f''(0), \\ \theta(0) = 1, \quad \phi(0) = 1, \\ \\ f'(\infty) = \lambda, \quad g(\infty) = 0, \quad \theta(\infty) \to 0, \\ \phi(\infty) \to 0, \quad \eta \to \infty\end{array}\right\} \tag{13}$$

where Sc is the Schmidt number, γ is the chemical reaction, K is the vortex viscosity, η is the similarity variable, λ is the velocity ratio parameter, Pr is the Prandtl number, D_f and S_r are the Dufour and Soret numbers, S is the suction injection parameter, and μ is the dynamic viscosity.

Here, different flow parameters are defined as

$$\left.\begin{array}{c}K = \frac{\kappa}{\mu}, \quad Sc = \frac{\nu}{D_e}, \quad \lambda = \frac{b}{a}, \quad \alpha = \frac{c}{a}, \\ \gamma = \frac{R_1}{a}, \quad \Pr = \nu/\alpha_m, \\ D_f = \frac{D_e k_T}{C_s C_p}\frac{(C_w - C_\infty)}{(T_w - T_\infty)\upsilon}, \quad Sr = \frac{D_e k_T}{T_m \upsilon}\frac{(T_w - T_\infty)}{(C_w - C_\infty)}\end{array}\right\} \tag{14}$$

The parameter of physical interest skin friction coefficient C_{fx}, the local Nusselt number Nu_x and the Sherwood number Sh are defined as

$$\left.\begin{array}{c}C_{fx} = \frac{\tau_w}{\rho U_w^2/2}, \quad Nu_x = \frac{xq_w}{k(T_w - T_\infty)}, \\ Sh = \frac{xj_w}{De(C_w - C_\infty)} \\ \tau_w = ((\mu + k)\frac{\partial u}{\partial y} + kN)_{y=0}, \quad q_w = -k\left(\frac{\partial T}{\partial y}\right)_{y=0}, \\ j_w = -D_e\left(\frac{\partial C}{\partial y}\right)_{y=0}\end{array}\right\} \tag{15}$$

From (9) and (16) we have

$$\left.\begin{aligned} C_{fx} R_{e_x}^{\frac{1}{2}} &= (1+K) f''(0) + K g(0) \\ Nu_x/\mathrm{Re}_x^{1/2} &= -\theta'(0),\ Sh/\mathrm{Re}_x^{1/2} = -\phi'(0). \end{aligned}\right\} \quad (16)$$

3 Analytical Solutions

We have used HAM to solve (9)–(12) with boundary conditions (13) analytically. Here we describe the procedure briefly.

First, we choose initial guesses for velocity (f_0) , microrotation (g_0),concentration (ϕ_0), and temperature (θ_0) with the help of set of base function that are defined as follows:

$$f_0(\eta) = S + \lambda * \eta + (\lambda - 1) * (\exp[-\eta] - 1), \quad (17)$$

$$g_0(\eta) = -N_0 * f_0^{*''}(0) \exp[-\eta], \quad (18)$$

$$\phi_0(\eta) = \exp(-\eta), \quad (19)$$

$$\theta_0(\eta) = \exp(-\eta), \quad (20)$$

Then we defined the auxiliary linear operators for Eqs. (9)–(12).

$$\left.\begin{aligned} L_f &= \frac{d^3 f}{d\eta^3} - \frac{df}{d\eta}, \quad L_g = \frac{d^2 g}{d\eta^2} - g, \\ L_\phi &= \frac{d^2\phi}{d\eta^2} - \phi, \quad L_\theta = \frac{d^2\theta}{d\eta^2} - \theta, \end{aligned}\right\} \quad (21)$$

Then we define zeroth-order deformation. After that we differentiate zero-order equations m times with respect to p and then divided by p! By choosing $p = 0$, we get mth order deformation. Finally, we get general solutions that are given below:

$$f_m(\eta) = f_m^*(\eta) + C_1 + C_2 \exp(\eta) + C_3 \exp(-\eta), \quad (22)$$

$$g_m(\eta) = g_m^*(\eta) + C_4 \exp(\eta) + C_5 \exp(-\eta), \quad (23)$$

$$\phi_m(\eta) = \phi_m^*(\eta) + C_6 \exp(\eta) + C_7 \exp(-\eta), \quad (24)$$

$$\theta_m(\eta) = \theta_m^*(\eta) + C_8 \exp(\eta) + C_9 \exp(-\eta). \quad (25)$$

where $f_m^*(\eta), g_m^*(\eta), \phi_m^*(\eta), \theta_m^*(\eta)$ are the particular solutions of the equations, where the values of constants from C_1 to C_9 can be found using boundary conditions.

4 Convergence of the Series Solutions

It is commonly known that a nonlinear problem can be approximated more efficiently through the choice of an appropriate set of base functions and to certify the convergence using HAM (Homotopy Analysis Method). The rate of approximation and convergence criteria depends upon the value of nonzero auxiliary parameter $\hbar$, usually called a convergence control parameter. Here for the case of velocity, microrotation, concentration, and temperature profile, $\hbar$ is represented as $\hbar_f$, $\hbar_g$, $\hbar_\varphi$, and $\hbar_\theta$, respectively. These parameters help in the adjustment and control the radius of convergence of the series solutions. The range of admissible values of $\hbar_f$, $\hbar_g$, $\hbar_\varphi$, and $\hbar_\theta$ can be computed by displaying the $\hbar-$ curves of the functions $f''(0), g'(0), \varphi'(0), \theta'(0)$. Figures 1, 2, 3, and 4 are drawn at 15th order of approximation for the fixed values of other parameters as $K = 1 = \lambda$, $\alpha = 0.5$, $D_f = 0.1$, $\gamma = 1 = \Pr$, $Sr = 1 = Sc$, $S = 0$, and the admissible values of $f''(0), g'(0), \varphi'(0), \theta'(0)$ lie in the range $-0.5 \leq \hbar_f \leq -0.2$, $-0.6 \leq \hbar_g \leq -0.3$, $-0.5 \leq \hbar_\phi \leq -0.2$, $-1.25 \leq \hbar_\theta \leq -0.6$, respectively.

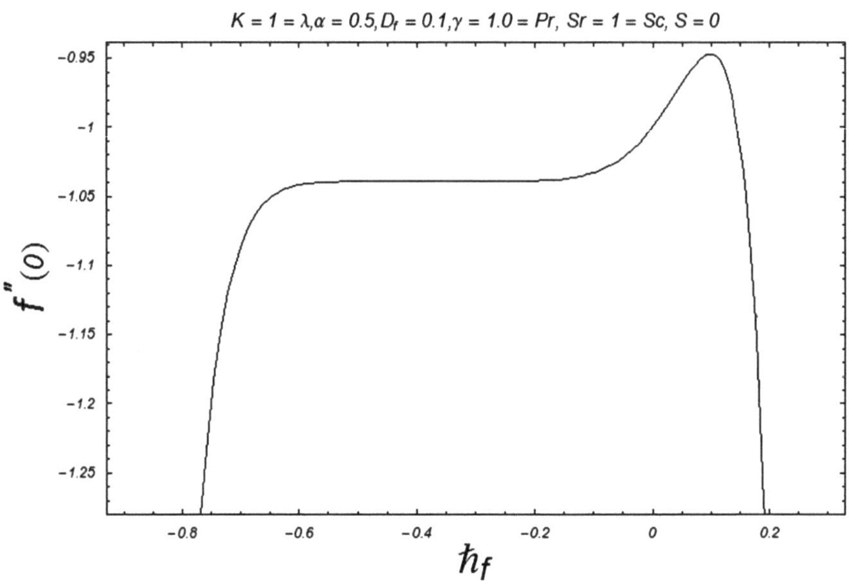

Fig. 1 $\hbar_f$ curve for $f'(0)$

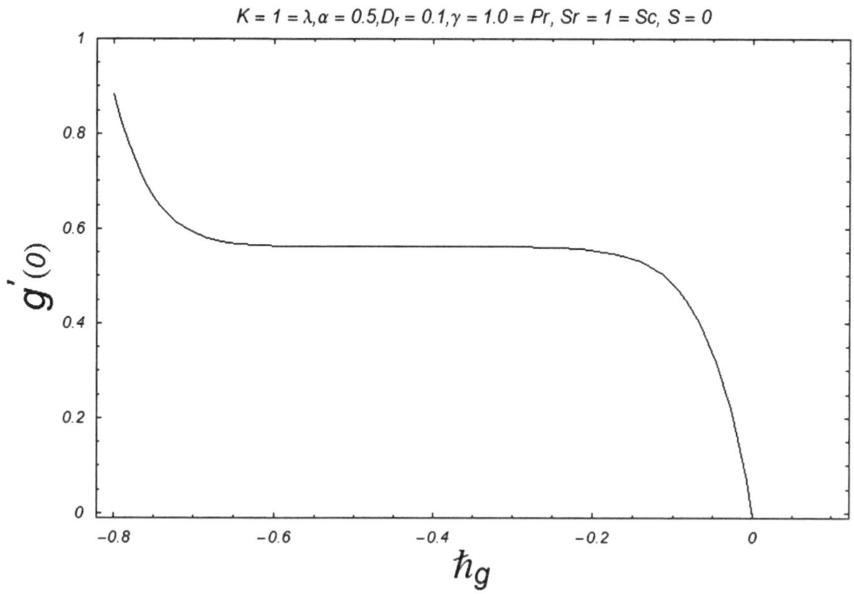

Fig. 2 $\hbar_g$ curve for $g'(0)$

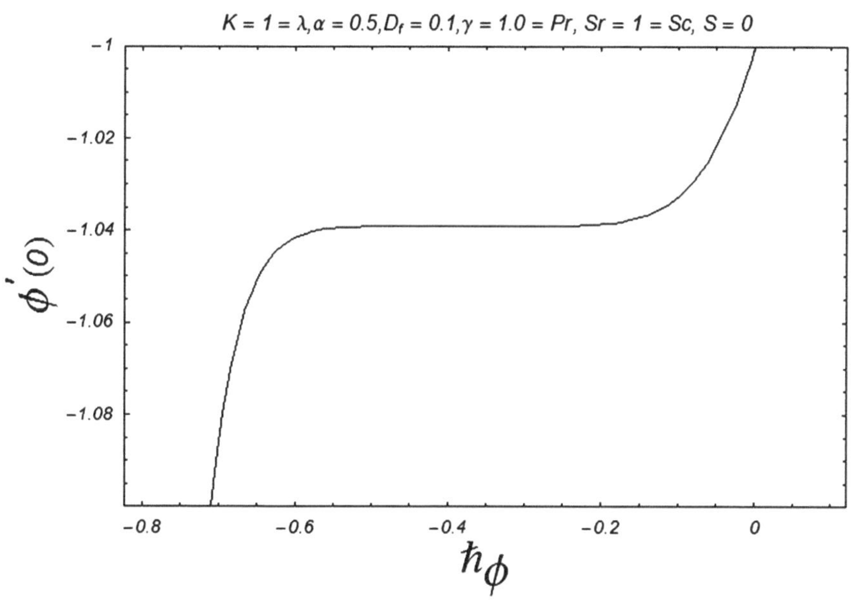

Fig. 3 $\hbar_\phi$ curve for $\phi'(0)$

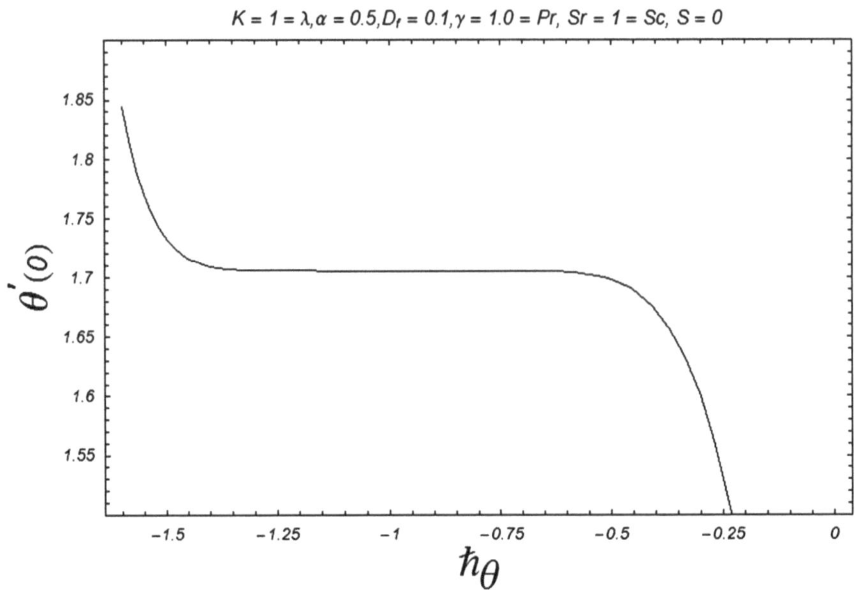

Fig. 4 $\hbar_\theta$ curve for $\theta'(0)$

5 Results and Discussion

Homotopy analysis method (HAM) is applied to solve the equation from (9) to (13) analytically using Mathematica 9 software. We investigate the influences of different parameters on concentration and temperature distribution to comprehend the physical phases of the problem. These embedded flow parameters are α (unsteady parameter), Sc (Schmidt number), γ (chemical reaction parameter), D_f (Dufour number), and S_r (Soret number). Figures 5, 6, 7, 8, and 9 show the effects of different flow parameters on concentration profiles, whereas Figs. 10, 11, and 12 indicate the influence of involved parameters on temperature profiles. Figure 5 shows the influence unsteady parameter α on concentration profiles. It is observed that concentration distribution increases by increasing the α for fixed values of other parameters, whereas Schmidt number shows opposite behavior on the concentration velocity which is displayed in Fig. 6 due to decrease in boundary layer thickness. Physically increase in Sc means that decrease in molecular diffusivity D, so as a result concentration distribution decreases.

The effects γ on concentration profiles are sketched in Fig. 7 which shows that as we increase the values of γ, the $\phi(\eta)$ increases. It means that the concentration of the diffusing species increases during chemical reaction. The Dufour and Soret numbers have increasing effects as shown in Figs. 8 and 9. With the increase of these parameters the concentration velocity for Soret number increases more rapidly as compared to Dufour number. It clears that diffusive species boost the concentration in

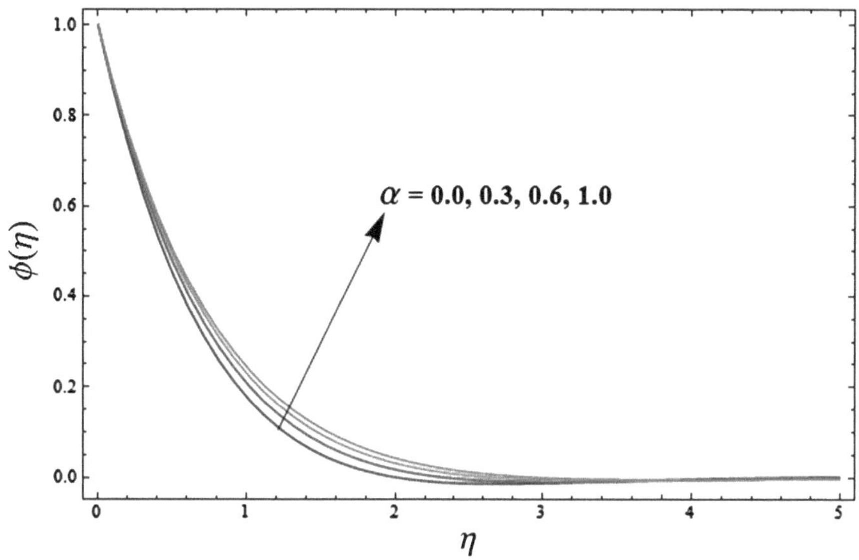

Fig. 5 Influence of α on concentration profiles when $\lambda = 2, Sc = 2, Sr = 1, D_f = 1, S = 2$

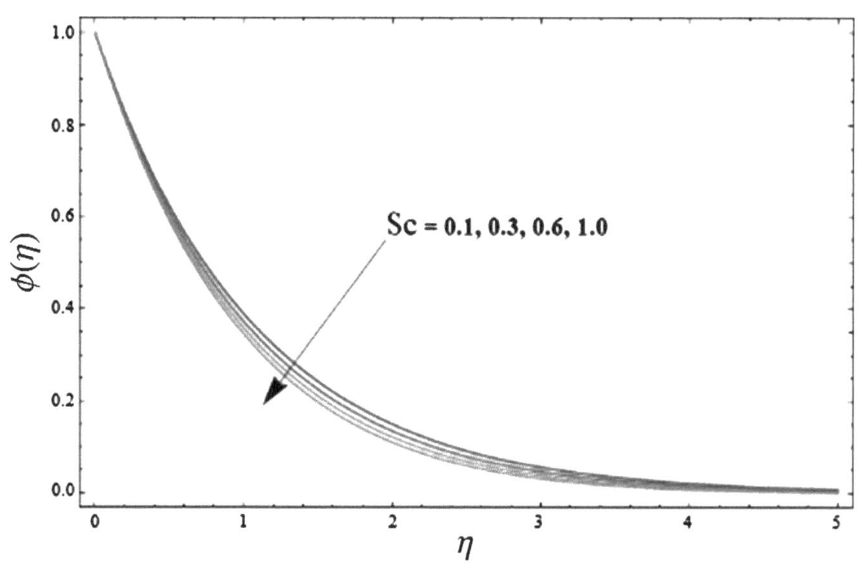

Fig. 6 Influence of Sc on concentration profiles when $\lambda = 2, \gamma = 0.3, Sr = 1, D_f = 0.2, S = 1, \alpha = 0.5$

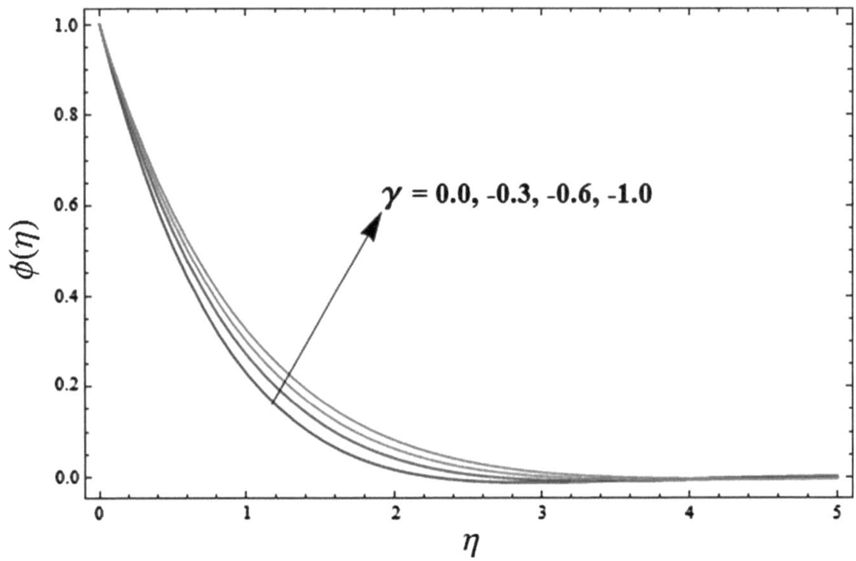

Fig. 7 Influence of γ on concentration profiles when $\lambda = 0.3$, $Sc = 3$, $Sr = 1$, $D_f = 2$, $S = 1$, $\alpha = 0.5$

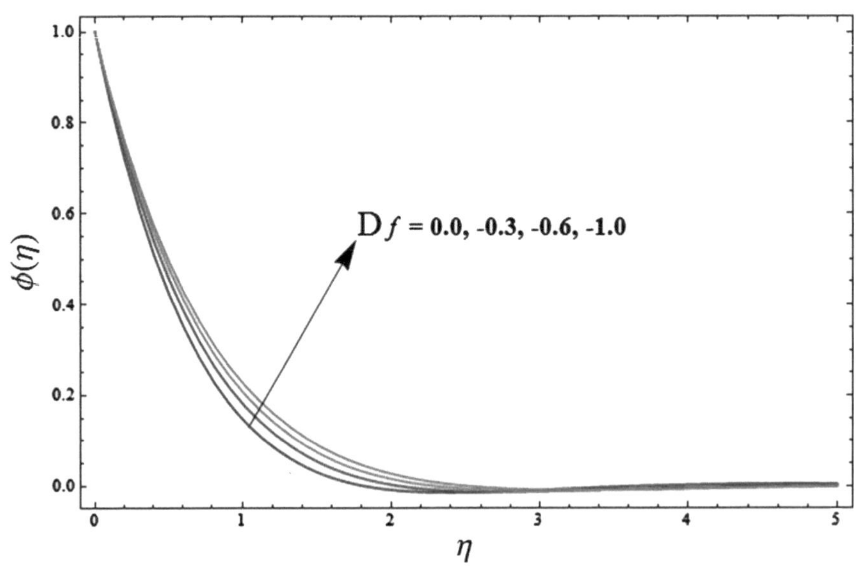

Fig. 8 Influence of D_f on concentration profiles when $\lambda = 0.3$, $Sc = 3$, $Sr = 1$, $\gamma = 2$, $S = 1$, $\alpha = 0.5$

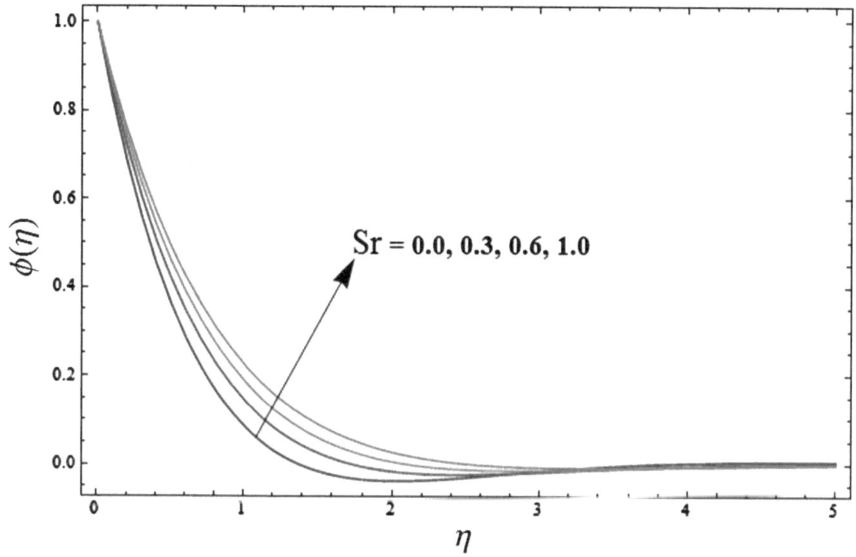

Fig. 9 Influence of *Sr* on concentration profiles when $\lambda = 0.3, Sc = 3, \gamma = 2, D_f = 1, S = 1, \alpha = 0.5$

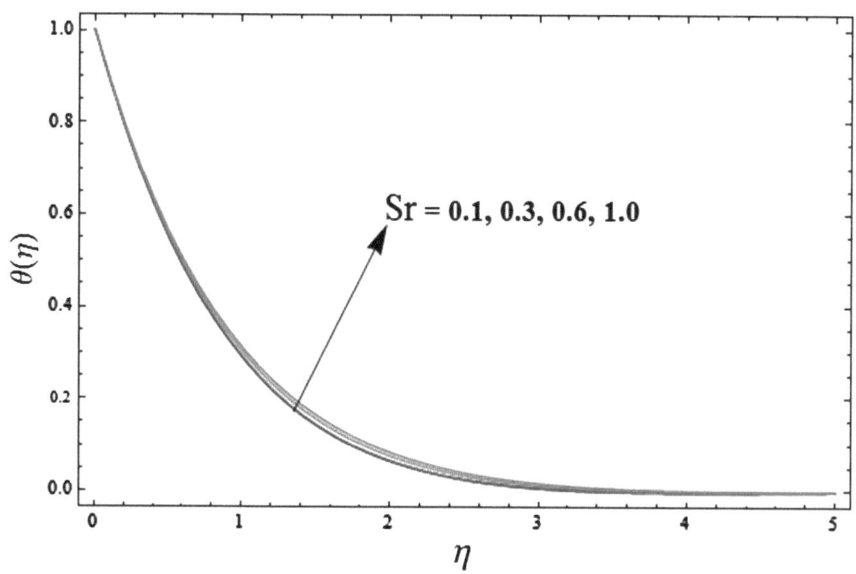

Fig. 10 Influence of *Sr* on temperature profiles when $\lambda = 2, \Pr = 2, D_f = 1, S = 1, \alpha = 0.5$

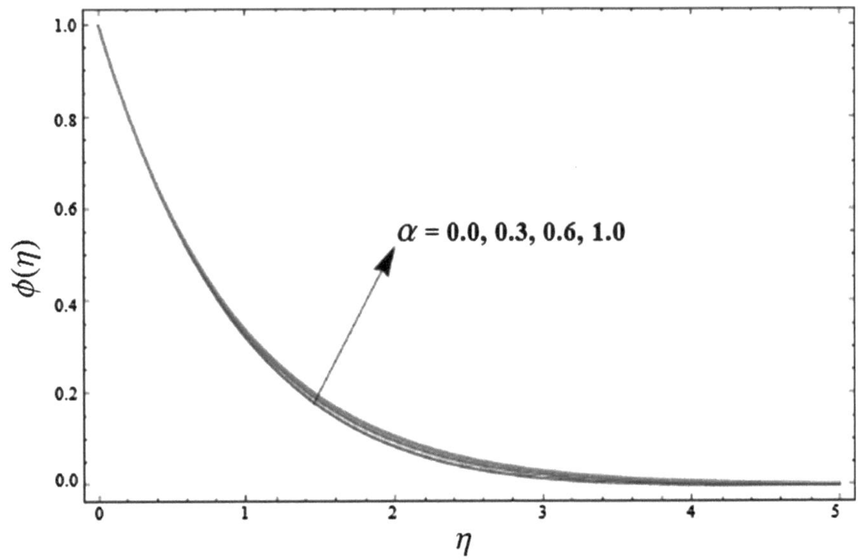

Fig. 11 Influence of α on temperature profiles when $\lambda = 2$, $Sc = 0.2$, $Sr = 1$, $D_f = 0.3$, $S = 1$

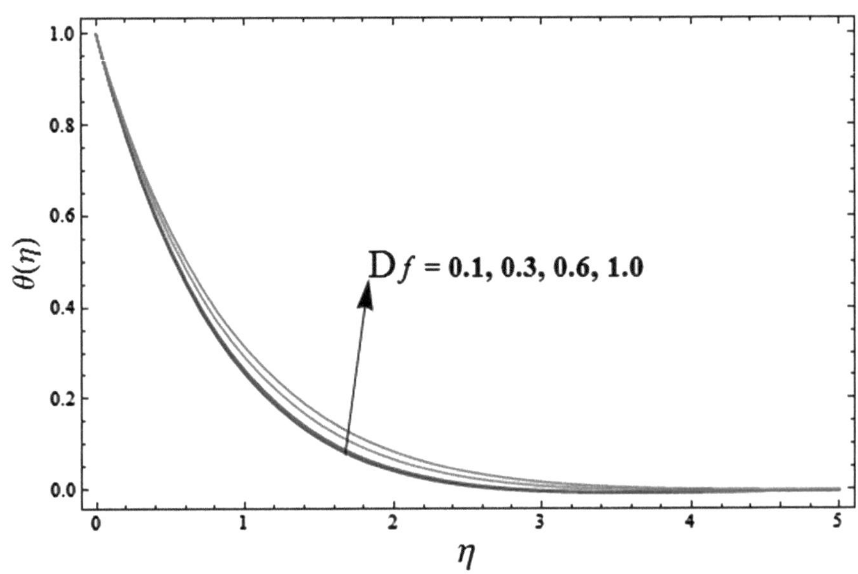

Fig. 12 Influence of D_f on temperature profiles when $\lambda = 2$, $Sr = 1$, $\Pr = 2$, $S = 1$, $\alpha = 0.5$

Table 1 Numerical values of skin friction coefficient $\mathrm{Re}_x^{1/2} C_f$ when $K = 0, 1$, at $S = 0$, $\alpha = 0, N_0 = 0$

λ	Mahapatra et al. [4]	$K = 0$ (Newtonian fluid)		$K = 1$	
		Nazar et al. [5]	Present result	Nazar et al. [5]	Present result
0.01	–	−0.9980	−0.9980	−1.3653	−1.3653
0.02	–	−0.9958	−0.9958	−1.3622	−1.3622
0.05	–	−0.9876	−0.9876	−1.3512	−1.3512
0.10	−0.9694	−0.9694	−0.9694	−1.3268	−1.3268
0.20	−0.9181	−0.9181	−0.9181	−1.2579	−1.2579
0.50	−0.6673	−0.6673	−0.6673	−0.9175	−0.9175
1.00	–	0.0000	0.0000	0.0000	0.0000

Table 2 Numerical values of skin friction coefficient $\mathrm{Re}_x^{1/2} C_f$ when $K = 0, 1$ at $S = 0$, $\alpha = 0, N_0 = 1/2$

λ	Mahapatra et al. [4]	$K = 0$ (Newtonian fluid)		$K = 1$	
		Nazar et al. [5]	Present result	Nazar et al. [5]	Present result
0.01	–	−0.9980	−0.9980	−1.2224	−1.2224
0.02	–	−0.9958	−0.9958	−1.2196	−1.2196
0.05	–	−0.9876	−0.9876	−1.2095	−1.2095
0.10	−0.9694	−0.9694	−0.9694	−1.1872	−1.1872
0.20	−0.9181	−0.9181	−0.9181	−1.1244	−1.1244
0.50	−0.6673	−0.6673	−0.6673	−0.8172	−0.8172
1.00	–	0.0000	0.0000	0.0000	0.0000

the flow field with Soret and Dufour effect. Figures 10, 11, and 12 are prepared to see the influence of Sr, α and D_f on temperature profiles. From these figures, it is obvious that S_r and α have slow behavior on the temperature velocity but with the increase of D_f, the temperature profiles enhanced more as compared to S_r and α. The present results for $S = 0$, $\alpha = 0$ are compared with the Ref. [4]. The validation of HAM results are in good agreement with the results presented in [5] when $S = 0$, $\alpha = 0$. Such comparisons are shown in Tables 1 and 2 when $N_0 = 0$ and $N_0 = 1/2$. This comparison is provided in good agreement.

6 Conclusions

The current study explores the effects of Soret and Dufour on the micropolar fluid toward a permeable stretching sheet for unsteady stagnation point flow case. The nonlinear ordinary differential equations (9)–(12) with boundary condition (13) are solved analytically by homotopy analysis method using Mathematica 9. The results for different flow parameters of interest are shown. So the following are the main points drawn from this study:

- $\hbar$—curves show the convergence of the series solution.
- The concentration profiles increase rapidly, whereas temperature profiles increase slowly with the increasing values of unsteady parameter α.
- Temperature velocity increases slowly with the increase of Soret number but for concentration distribution increases fast.
- With the increase of chemical reaction γ, concentration distribution increases while for Schmidt number Sc reduces gradually.
- Dufour number increases the concentration and temperature distributions.

Acknowledgments The first author is grateful to the University Technology PETRONAS for financial support.

References

1. Eringen, C.: Theory of micropolar fluids. J. Math. Mech. **16**, 1–18 (1966)
2. Joneidi, A., Ganji, D.D., Babaelahi, M.: Micropolar flow in a porous channel with high mass transfer. Int. Commun. Heat Mass Trans. **36**, 1082–1088 (2009)
3. Ashraf, M., Kamal, M.A., Syed, K.S.: Numerical study of asymmetric laminar flow of micropolar fluids in a porous channel. Comput. Fluids. **38**, 1895–1902 (2009)
4. Mahapatra, T.R., Gupta, A.S.: Heat transfer in stagnation-point flow towards a stretching sheet. Heat Mass Trans. **38**, 517–521 (2002)
5. Nazar, R., Amin, N., Filip, D., Pop, I.: Stagnation point flow of a micropolar fluid towards a stretching sheet. Int. J. Nonlinear Mech. **39**, 1227–1235 (2004)
6. Angel, M., Takhar, H.S., Pop, I.: Dufour and Soret effects on free convection boundary layer over a vertical surface embedded in a porous medium, Studia univer. Bolyai, Mathematica XLV, pp. 11–21 (2000)
7. Cheng, C.Y.: Soret and Dufour effects on heat and mass transfer by natural convection from a vertical truncated cone in a fluid-saturated porous medium with variable wall temperature and concentration. Int. Commun. Heat Mass Transf. **37**, 1031–1035 (2010)
8. Pal, D., Chatterjee, S.: Mixed convection magnetohydrodynamic heat and mass transfer past a stretching surface in a micropolar fluid-saturated porous medium under the influence of Ohmic heating, Soret and Dufour effects. Commun. Nonlinear Sci. Numer. Simulat. **16**, 1329–1346 (2011)
9. Pal, D., Mondal, H.: Influence of thermophoresis and Soret-Dufour on magnetohydrodynamic heat and mass transfer over a non-isothermal wedge with thermal radiation and Ohmic dissipation. J. Magn. Magn. Mater. **331**, 250–255 (2013)
10. Pal, D., Mondal, H.: Effects of Soret Dufour, chemical reaction and thermal radiation on MHD non-Darcy unsteady mixed convective heat and mass transfer over a stretching sheet. Commun. Nonlinear Sci. Numer. Simulat. **16**, 1942–1958 (2011)
11. Pal, D., Mondal, H.: Influence of Soret and Dufour on MHD buoyancy-driven heat and mass transfer over a stretching sheet in porous media with temperature-dependent viscosity. Nucl. Eng. Des. **256**, 350–357 (2013)

12. Pal, D., Mondal, H.: Soret and Dufour effects on MHD non-Darcian mixed convection heat and mass transfer over a stretching sheet with nonuniform heat source/sink. Phys B: Condens Matter. **407**, 642–651 (2012)
13. Liao, S.J.: A general approach to get series solution of nonsimilarity boundary layer flows. Commun. Nonlinear Sci. Numer. Simulat. **14**, 2144–2159 (2009)
14. Liao, S.J.: Notes on the homotopy analysis method: Some definitions and theorems. Commun. Nonlinear Sci. Numer. Simulat. **14**, 983–997 (2009)
15. Peddieson, J., Mc, R.P.: Boundary layer theory for micropolar fluid. Recent Adv. Eng. Sci. **5**, 405–426 (2009)

One-Step Implicit Hybrid Method for Solving Semi-explicit Index-1 Differential Algebraic Equations

Khoo Kai Wen and Zanariah Abdul Majid

Abstract In this paper, a self-starting one-step implicit hybrid method is proposed to solve semi-explicit index-1 differential algebraic equations (DAEs). The proposed method is formulated using Lagrange interpolating polynomial. The proposed method will compute the solutions of differential algebraic equations using constant step size. Implementation of the method involved Newton's iteration in order to solve semi-explicit index-1 differential algebraic equations. Numerical examples are shown in order to present the applicability of the proposed method when solving the semi-explicit index-1 differential algebraic equations. The results of proposed method show better results compared to existing methods when solving semi-explicit index-1 differential algebraic equations.

Keywords Differential algebraic equations · Hybrid method · Semi-explicit · Self-starting · Constant step size

1 Introduction

This research focused on study numerical solutions for semi-explicit index-1 differential algebraic equations (DAEs). Semi-explicit index-1 DAEs consist of a system of ordinary differential equations with algebraic constrains, and it can be written in following form:

$$\begin{aligned} y' &= f(y, z) \quad y(x_0) = y_0 \\ 0 &= g(y, z) \quad z(z_0) = z_0 \end{aligned} \tag{1}$$

where $g_z = \frac{\partial g}{\partial z}$ is nonsingular in a neighborhood of solution. The unknowns y and z represent differential variable and algebraic variable, respectively.

K. Kai Wen (✉) · Z. Abdul Majid
Institute for Mathematical Research, Universiti Putra Malaysia, Serdang, Selangor, Malaysia
e-mail: kkw1025@hotmail.com

A. Kılıçman et al. (eds.), *Recent Advances in Mathematical Sciences*,
DOI 10.1007/978-981-10-0519-0_6

One-step method in numerical analysis indicated that the method used information from only one of the previous points to determine the approximation at the next point. By referring to Butcher (2002), hybrid method can be formed in this way that needs two predictors, one to obtain an approximation to the solution at the off-step point and a second to give a first approximation to $y(x_n)$, from which f_n is calculated.

In literature, Gear (1971) had introduced the ideas of applying numerical method of ODEs in order to solve DAEs. Numerical solutions of ODE methods for solving DAEs are derived by Gear and Pezold (1984). They concluded that the level of difficulties to solve DAEs depends on the index of DAEs. By referring to Petzold (1982), they had discussed the difficulties in solving DAEs, and stated that the basic algorithms will work for lower index DAEs. Recently, several methods have been proposed to solve semi-explicit index-1 DAEs, such as [1], 2-points and 3-points BBDF [2], and L-stable extended BBDF [3]. Hybrid method is a method of modified linear multistep formulae which incorporate a function evaluation at an off-step point [4]. There are many researchers who had tried to use hybrid method to solve stiff ODEs, such as in [5–7].

2 Formation of Method

Based on Fig. 1, one-step implicit hybrid method is formulated using Lagrange interpolating polynomial with one off-step point. It is a self-staring method and it will compute the approximate solutions using constant step size. The interpolating points are y_n, y_{n+1}, and $y_{n+\frac{5}{4}}$. The proposed method used the extended point $y_{n+\frac{5}{4}}$ which is $\frac{1}{4}h$ apart from y_{n+1} to compute the new value of y_{n+1}. The derivation of the implicit hybrid method is obtained by taking the first-order differential equation as follows:

$$y' = f(x, y) \tag{2}$$

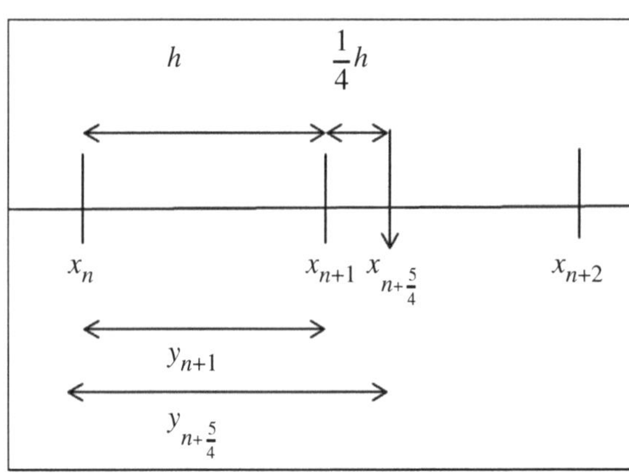

Fig. 1 One-step implicit hybrid method

The approximation value of y_{n+1} for the first equation can be obtained by integrating Eq. 2 from x_n to x_{n+1} with respect to x as shown as follows:

$$\int_{x_n}^{x_{n+1}} y' dx = \int_{x_n}^{x_{n+1}} f(x, y) dx$$

or

$$y(x_{n+1}) = y(x_n) + \int_{x_n}^{x_{n+1}} f(x, y) dx \tag{3}$$

Next, $y' = f(x, y)$ in Eq. 3 is replaced by Lagrange interpolation polynomial. The involved interpolation points are (x_n, f_n), (x_{n+1}, f_{n+1}), and $(x_{n+\frac{5}{4}}, f_{n+\frac{5}{4}})$, and the Lagrange interpolating polynomial is shown as follows:

$$\begin{aligned} P = & \frac{(x - x_{n+1})(x - x_{n+\frac{5}{4}})}{(x_n - x_{n+1})(x_n - x_{n+\frac{5}{4}})} f_n \\ & + \frac{(x - x_n)(x - x_{n+\frac{5}{4}})}{(x_{n+1} - x_n)(x_{n+1} - x_{n+\frac{5}{4}})} f_{n+1} \\ & + \frac{(x - x_n)(x - x_{n+1})}{(x_{n+\frac{5}{4}} - x_n)(x_{n+\frac{5}{4}} - x_{n+1})} f_{n+\frac{5}{4}} \end{aligned}$$

Then, by taking $s = \frac{x - x_{n+\frac{5}{4}}}{h}$, $dx = hds$ and replacing into Eq. 3. Let $x = x_n$,

$$s = \frac{x_n - x_{n+\frac{5}{4}}}{h}, \quad s = \frac{-\frac{5}{4}h}{h}$$

So, $s = -\frac{5}{4}$, then, taking $x = x_{n+1}$,

$$s = \frac{x_{n+1} - x_{n+\frac{5}{4}}}{h}, \quad s = \frac{-\frac{1}{4}h}{h}$$

So $s = -\frac{1}{4}$, hence, limit of integration is from $-\frac{5}{4}$ to $-\frac{1}{4}$. Hence, the first equation of one-step hybrid method will be obtained as shown as follows:

$$y_{n+1} = y_n + \frac{11}{30} h f_n + \frac{7}{6} h f_{n+1} - \frac{8}{15} h f_{n+\frac{5}{4}} \tag{4}$$

The same process is repeated in order to obtain the second equation of the formula where integrating Eq. 2 from x_n to $x_{n+\frac{5}{4}}$ with respect to x,

$$\int_{x_n}^{x_{n+\frac{5}{4}}} y' dx = \int_{x_n}^{x_{n+\frac{5}{4}}} f(x, y) dx$$

or

$$y(x_{n+\frac{5}{4}}) = y(x_n) + \int_{x_n}^{x_{n+\frac{5}{4}}} f(x, y)dx \tag{5}$$

Then, change the limit of integration from $-\frac{5}{4}$ to 0. Thus, the second equation of the method can be obtained as follows:

$$y_{n+\frac{5}{4}} = y_n + \frac{35}{96}hf_n + \frac{125}{96}hf_{n+1} - \frac{5}{12}hf_{n+\frac{5}{4}} \tag{6}$$

Hence, the one-step implicit hybrid method is obtained and shown as below:

$$\begin{aligned} y_{n+1} &= y_n + \frac{11}{30}hf_n + \frac{7}{6}hf_{n+1} - \frac{8}{15}hf_{n+\frac{5}{4}} \\ y_{n+\frac{5}{4}} &= y_n + \frac{35}{96}hf_n + \frac{125}{96}hf_{n+1} - \frac{5}{12}hf_{n+\frac{5}{4}} \end{aligned} \tag{7}$$

3 Analysis of the Method

By referring to Jator (2010), the order of this method is identified by applying the constant coefficient, C_p, $p = 0, 1, 2, \ldots,$ the formulae can be defined as follows:

$$\begin{aligned} C_0 &= \sum_{j=0}^{k} \alpha_j \\ C_1 &= \sum_{j=1}^{k} j\alpha_j - \sum_{j=1}^{k} \beta_j - \sum_{j=1}^{1} \beta_{vj} \\ &\vdots \\ C_p &= \frac{1}{p!}\left[\sum_{j=1}^{k} j^p \alpha_j - p\left(\sum_{j=1}^{k} j^{p-1}\beta_j + \sum_{j=1}^{1} v_j^{p-1}\beta_{vj}\right)\right] \end{aligned} \tag{8}$$

Hence, the order and error constant of the method will be computed using Eq. 8 as shown as follows:

For $p = 0, 1, 2, 3, 4$:

$$C_0 = \sum_{j=0}^{k} \alpha_j = \begin{pmatrix} -1 \\ -1 \end{pmatrix} + \begin{pmatrix} 1 \\ 0 \end{pmatrix} + \begin{pmatrix} 0 \\ 1 \end{pmatrix} = \begin{pmatrix} 0 \\ 0 \end{pmatrix}$$

$$
\begin{aligned}
C_1 &= \sum_{j=1}^{k} j\alpha_j - \sum_{j=1}^{k} \beta_j - \sum_{j=1}^{1} \beta_{vj} \\
&= \begin{pmatrix} 1 \\ 0 \end{pmatrix} + \frac{5}{4}\begin{pmatrix} 0 \\ 1 \end{pmatrix} - \begin{pmatrix} \frac{11}{30} \\ \frac{35}{96} \end{pmatrix} - \begin{pmatrix} \frac{7}{6} \\ \frac{125}{96} \end{pmatrix} - \begin{pmatrix} -\frac{8}{15} \\ -\frac{5}{12} \end{pmatrix} = \begin{pmatrix} 0 \\ 0 \end{pmatrix} \\
C_2 &= \frac{1}{2!}\left[\sum_{j=1}^{k} j^2\alpha_j - 2\left(\sum_{j=1}^{k} j^{2-1}\beta_j + \sum_{j=1}^{1} v_j^{2-1}\beta_{vj}\right)\right] \\
&= \frac{1}{2}\left[\begin{pmatrix} 1 \\ 0 \end{pmatrix} + \begin{pmatrix} \frac{5}{4} \end{pmatrix}^2 \begin{pmatrix} 0 \\ 1 \end{pmatrix} - 2\left(\begin{pmatrix} \frac{7}{6} \\ \frac{125}{96} \end{pmatrix} + \begin{pmatrix} \frac{5}{4} \end{pmatrix}\begin{pmatrix} -\frac{8}{15} \\ -\frac{5}{12} \end{pmatrix}\right)\right] = \begin{pmatrix} 0 \\ 0 \end{pmatrix} \\
C_3 &= \frac{1}{3!}\left[\sum_{j=1}^{k} j^3\alpha_j - 3\left(\sum_{j=1}^{k} j^{3-1}\beta_j + \sum_{j=1}^{1} v_j^{3-1}\beta_{vj}\right)\right] \\
&= \frac{1}{6}\left[\begin{pmatrix} 1 \\ 0 \end{pmatrix} + \begin{pmatrix} \frac{5}{4} \end{pmatrix}^3 \begin{pmatrix} 0 \\ 1 \end{pmatrix} - 3\left(\begin{pmatrix} \frac{7}{6} \\ \frac{125}{96} \end{pmatrix} + \begin{pmatrix} \frac{5}{4} \end{pmatrix}^2\begin{pmatrix} -\frac{8}{15} \\ -\frac{5}{12} \end{pmatrix}\right)\right] = \begin{pmatrix} 0 \\ 0 \end{pmatrix} \\
C_4 &= \frac{1}{4!}\left[\sum_{j=1}^{k} j^4\alpha_j - 4\left(\sum_{j=1}^{k} j^{4-1}\beta_j + \sum_{j=1}^{1} v_j^{4-1}\beta_{vj}\right)\right] \\
&= \frac{1}{24}\left[\begin{pmatrix} 1 \\ 0 \end{pmatrix} + \begin{pmatrix} \frac{5}{4} \end{pmatrix}^4 \begin{pmatrix} 0 \\ 1 \end{pmatrix} - 4\left(\begin{pmatrix} \frac{7}{6} \\ \frac{125}{96} \end{pmatrix} + \begin{pmatrix} \frac{5}{4} \end{pmatrix}^3\begin{pmatrix} -\frac{8}{15} \\ -\frac{5}{12} \end{pmatrix}\right)\right] = \begin{pmatrix} \frac{1}{48} \\ \frac{125}{6144} \end{pmatrix} \neq \begin{pmatrix} 0 \\ 0 \end{pmatrix}
\end{aligned}
$$

A method is order of p if $C_0 = C_1 = \cdots = C_p = 0$ and $C_{p+1} \neq 0$ is the error constant of the method. Since $C_4 \neq 0$, hence the order of this method is three and the error constant is

$$
C_{p+1} = C_4 = \begin{pmatrix} \frac{1}{48} \\ \frac{125}{6144} \end{pmatrix}.
$$

The stability of one-step implicit hybrid method Eq. 7 can be determined when the method is applied to the test equation:

$$
y' = f = \lambda y \tag{9}
$$

The characteristic polynomial of the method can be obtained as follows:

$$
\left(-1 - \frac{1}{2}\bar{h} - \frac{5}{96}\bar{h}^2\right)t + \left(1 - \frac{3}{4}\bar{h} + \frac{5}{24}\bar{h}^2\right)t^2 = 0 \tag{10}
$$

where $\bar{h} = h\lambda$, stability region of the method can be determined using software MAPLE with Eq. 10. Hence, the stability is shown as in Fig. 2.
By referring to the definition, the method Eq. 7 can be considered as A-stable since the stability region of Eq. 7 covered the left-hand plane. A-stable method is suitable to solve stiff equations.

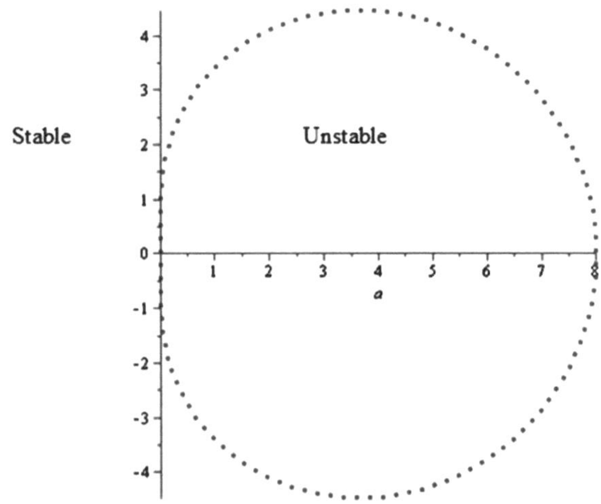

Fig. 2 Stability region of one-step implicit hybrid method

4 Implementation of Method

One-step implicit hybrid method is implemented with Newton's iteration in order to solve semi-explicit DAE. This method is a self-starting method with predictor–corrector scheme. Prediction of pointsy_{n+1} and $y_{n+\frac{5}{4}}$ can be done using the initial value with the predictor formulae. The predicted value will then be used in the corrector formulae to compute the approximate value of y_{n+1}.
Define

$$F_1 = y_{n+1} - y_n - \frac{7}{30}hf_n - \frac{7}{6}hf_{n+1} + \frac{8}{15}hf_{n+\frac{5}{4}}$$

$$F_{\frac{5}{4}} = y_{n+\frac{5}{4}} - y_n - \frac{35}{96}hf_n - \frac{125}{96}hf_{n+1} + \frac{5}{12}hf_{n+\frac{5}{4}}$$

$$g_1 = g(y_{n+1}, z_{n+1})$$

$$g_{\frac{5}{4}} = g(y_{n+\frac{5}{4}}, z_{n+\frac{5}{4}}) \tag{11}$$

The formula of Newton's iteration is shown as follows:

$$y^{(i+1)} = y^{(i)} - J^{-1}(y^{(i)})f(y^{(i)}) \tag{12}$$

Then, substituting Eq. 11 into Eq. 12,

$$\begin{bmatrix} y_{n+j}^{(i+1)} \\ z_{n+j}^{(i+1)} \end{bmatrix} = \begin{bmatrix} y_{n+j}^{(i)} \\ z_{n+j}^{(i)} \end{bmatrix} - \begin{bmatrix} \frac{\partial F_j^{(i)}}{\partial y_{n+j}^{(i)}} & \frac{\partial F_j^{(i)}}{\partial z_{n+j}^{(i)}} \\ \frac{\partial g_j^{(i)}}{\partial y_{n+j}^{(i)}} & \frac{\partial g_j^{(i)}}{\partial z_{n+j}^{(i)}} \end{bmatrix}^{-1} \times \begin{bmatrix} F_j^{(i+1)} \\ g_j^{(i+1)} \end{bmatrix} \tag{13}$$

let

$$e_{n+j}^{(i+1)} = y_{n+j}^{(i+1)} - y_{n+j}^{(i)}$$
$$\hat{e}_{n+j}^{(i+1)} = z_{n+j}^{(i+1)} - z_{n+j}^{(i)} \quad \text{where } j = 1, \frac{5}{4}.$$

Hence,

$$\underbrace{\begin{bmatrix} \frac{\partial F_1^{(i)}}{\partial y_{n+1}^{(i)}} & \frac{\partial F_1^{(i)}}{\partial y_{n+\frac{5}{4}}^{(i)}} & \frac{\partial F_1^{(i)}}{\partial z_{n+1}^{(i)}} & \frac{\partial F_1^{(i)}}{\partial z_{n+\frac{5}{4}}^{(i)}} \\ \frac{\partial F_{\frac{5}{4}}^{(i)}}{\partial y_{n+1}^{(i)}} & \frac{\partial F_{\frac{5}{4}}^{(i)}}{\partial y_{n+\frac{5}{4}}^{(i)}} & \frac{\partial F_{\frac{5}{4}}^{(i)}}{\partial z_{n+1}^{(i)}} & \frac{\partial F_{\frac{5}{4}}^{(i)}}{\partial z_{n+\frac{5}{4}}^{(i)}} \\ \frac{\partial g_1^{(i)}}{\partial y_{n+1}^{(i)}} & \frac{\partial g_1^{(i)}}{\partial y_{n+\frac{5}{4}}^{(i)}} & \frac{\partial g_1^{(i)}}{\partial z_{n+1}^{(i)}} & \frac{\partial g_1^{(i)}}{\partial z_{n+\frac{5}{4}}^{(i)}} \\ \frac{\partial g_{\frac{5}{4}}^{(i)}}{\partial y_{n+1}^{(i)}} & \frac{\partial g_{\frac{5}{4}}^{(i)}}{\partial y_{n+\frac{5}{4}}^{(i)}} & \frac{\partial g_{\frac{5}{4}}^{(i)}}{\partial z_{n+1}^{(i)}} & \frac{\partial g_{\frac{5}{4}}^{(i)}}{\partial z_{n+\frac{5}{4}}^{(i)}} \end{bmatrix}}_{Jacobian\, matrix} \times \begin{bmatrix} e_{n+1}^{(i+1)} \\ e_{n+\frac{5}{4}}^{(i+1)} \\ \hat{e}_{n+1}^{(i+1)} \\ \hat{e}_{n+\frac{5}{4}}^{(i+1)} \end{bmatrix} = -\begin{bmatrix} F_1^{(i)} \\ F_{\frac{5}{4}}^{(i)} \\ g_1^{(i)} \\ g_{\frac{5}{4}}^{(i)} \end{bmatrix} \tag{14}$$

By referring to Eq. 14, the Jacobian matrix is shown as follows:

$$J = \begin{bmatrix} 1 - \frac{7h}{6}\frac{\partial f_{n+1}}{\partial y_{n+1}} & \frac{8h}{15}\frac{\partial f_{n+1}}{\partial y_{n+\frac{5}{4}}} & -\frac{7h}{6}\frac{\partial f_{n+1}}{\partial z_{n+1}} & \frac{8h}{15}\frac{\partial f_{n+1}}{\partial z_{n+\frac{5}{4}}} \\ -\frac{125h}{96}\frac{\partial f_{n+\frac{5}{4}}}{\partial y_{n+1}} & 1 + \frac{5h}{12}\frac{\partial f_{n+\frac{5}{4}}}{\partial y_{n+\frac{5}{4}}} & -\frac{125h}{96}\frac{\partial f_{n+\frac{5}{4}}}{\partial z_{n+1}} & \frac{5h}{12}\frac{\partial f_{n+\frac{5}{4}}}{\partial z_{n+\frac{5}{4}}} \\ \frac{\partial g_1}{\partial y_{n+1}} & \frac{\partial g_1}{\partial y_{n+\frac{5}{4}}} & \frac{\partial g_1}{\partial z_{n+1}} & \frac{\partial g_1}{\partial z_{n+\frac{5}{4}}} \\ \frac{\partial g_{\frac{5}{4}}}{\partial y_{n+1}} & \frac{\partial g_{\frac{5}{4}}}{\partial y_{n+\frac{5}{4}}} & \frac{\partial g_{\frac{5}{4}}}{\partial z_{n+1}} & \frac{\partial g_{\frac{5}{4}}}{\partial z_{n+\frac{5}{4}}} \end{bmatrix} \tag{15}$$

Thus, the values of y_{n+j} and z_{n+j} , where $j = 1, \frac{5}{4}$, will be obtained. The predicted values of z_{n+1} and $z_{n+\frac{5}{4}}$ can be computed from $g(x, y) = 0$ where $y_{n+1}^{(p)}$ and $y_{n+\frac{5}{4}}^{(p)}$ have been interpolated at the previous points. The absolute error of this solution is $|y_{n+j}^{(i+1)} - y_{n+j}^{(i)}| < TOL$ and $|z_{n+j}^{(i+1)} - z_{n+j}^{(i)}| < TOL$ where $TOL = 10^{-5}$.

5 Numerical Results

The following shows the test problems which are solved using one-step implicit hybrid method and compare with existing methods. The algorithm was written in C language.

Test problem 1:

$$\begin{aligned} y' &= f(y, z) = z & y(0) &= 1 \\ 0 &= g(y, z) = z^3 - y^2 & z(0) &= 1 \end{aligned} \quad 0 \leq x \leq 10$$

Exact solution:

$$y = \left(1 + \frac{x}{3}\right)^3$$
$$z = \left(1 + \frac{x}{3}\right)^2$$

Test problem 2:

$$\begin{aligned} y' &= f(x, y) = x\cos x - y + (1 + x)z \quad & y(0) = 1 \\ 0 &= g(x, y) = \sin x - z & z(0) = 0 \end{aligned} \qquad 0 \le x \le 10$$

Exact solution:

$$y = e^{-x} + x\sin x$$
$$z = \sin x$$

Notation:

h	Step size
1BDF	1-point sequential BDF method in Abasi et al. [2]
2BDF	2-point block BDF method in Abasi et al. [2]
3BDF	3-point block BDF method in Abasi et al. [2]
IHM	One-step implicit hybrid method
MAXE	Maximum error of the computed solution
Time	The execution time in second
TS	Total steps
TFC	Total function call

The error is defined in the following form:

$$error^{(i)} = Max(|y^{(i)}_{exact} - y^{(i)}_{approximate}|, |z^{(i)}_{exact} - z^{(i)}_{approximate}|)$$

and the maximum error is

$$MAXE = \max_{1 \le i \le N}(error^{(i)})$$

Tables 1 and 2 show the maximum error of the proposed method with three different step sizes. It is shown that the solution of IHM is more accurate than 1BDF, 2BDF, and 3BDF as the step size decreased. Besides, the execution time for IHM is more than 1BDF in test problem 1, but less than 1BDF in test problem 2. Moreover, the execution time for IHM is lesser than 2BDF and 3BDF for both of the test problems.

Table 1 Numerical results for test problem 1

h	Result				
	Method	Maxe	Time	TS	TFC
0.01	1BDF	2.7469e−1	1.8110e−3	–	–
	2BDF	1.9608e−3	2.1990e−3	–	–
	3BDF	2.0417e−3	2.3417e−3	–	–
	IHM	2.0189e−4	2.1990e−3	1000	1001
0.001	1BDF	2.7528e−2	1.9337e−2	–	–
	2BDF	1.9799e−5	4.4437e−2	–	–
	3BDF	2.0631e−5	4.8578e−2	–	–
	IHM	2.0137e−6	2.1922e−2	10,000	10,001
0.0001	1BDF	2.7533e−3	2.0087e−1	–	–
	2BDF	1.9798e−7	2.3496e−1	–	–
	3BDF	2.0636e−7	2.7629e−1	–	–
	IHM	1.9943e−8	2.1978e−1	100,000	100,001

Table 2 Numerical results for test problem 2

h	Result				
	Method	Maxe	Time	TS	TFC
0.01	1BDF	2.9999e−2	4.5121e−3	–	–
	2BDF	4.7146e−4	5.8129e−3	–	–
	3BDF	4.7815e−4	6.9001e−3	–	–
	IHM	1.3626e−7	1.3361e−3	1000	1001
0.001	1BDF	3.0062e−3	1.5927e−2	–	–
	2BDF	4.8149e−6	4.0054e−2	–	–
	3BDF	4.9186e−6	4.4136e−2	–	–
	IHM	1.3934e−10	1.2513e−2	10,000	10,001
0.0001	1BDF	3.0068e−4	1.4517e−1	–	–
	2BDF	4.8250e−8	2.1078e−1	–	–
	3BDF	4.9454e−8	2.3529e−1	–	–
	IHM	1.6230e−11	1.2586e−1	100,000	100,001

Meaning that the time duration for solving DAEs using one-step implicit hybrid method is faster than 2BDF and 3BDF. Hence, the proposed method which is one-step implicit hybrid method is suitable to solve semi-explicit index-1 DAEs.

References

1. El-Khateb, M.A., Hussien, H.S.: An optimization method for solving some differential algebraic equations. SIAM J. Sci. Numer. Simul. **4**, 1970–1977 (2009)
2. Abasi, N., Suleiman, M., Ibrahim, Z.B., Ismail, F.: 2-point and 3-point BBDF methods for solving semi-explicit index-1 DAEs. Appl. Math. Sci. **6**, 6679–6689 (2012)
3. Akinfenwa, O.A., Okunuga, S.A.: Solving semi-explicit index-1 DAE systems using L-stable extended block backward differentiation formula with constant coefficients. In: World Congress on Engineering (2013, in press)
4. Lambert, J.D.: Computational Methods in Ordinary Differential Equations. J.W. Arrowsmith Ltd., Bristol (1973)
5. Skwame, Y., Sunday, J., Ibijola, E.A.: L-stable block hybrid simpson's methods for numerical solution of initial value problems in stiff ordinary differential equations. Int. J. Pure Appl. Sci. Technol. **11**(2), 45–54 (2012)
6. Akinfenwa, O.A., Yao, N.M., Jator, S.N.: Implicit two step continuous hybrid block methods with four off-steps points for solving stiff ordinary differential equation. In: World Academy of Science, Engineering and Technology (2011, in press)
7. Jator, S.N.: On the hybrid method with three off-step points for initial value problems. Int. J. Math. Edu. Sci. Technol. **41**(1), 110–118 (2010)

An Artificial Intelligence Strategy for the Prediction of Wind Speed and Direction in Sarawak for Wind Energy Mapping

S.M. Lawan, W.A.W.Z. Abidin, S. Lawan and A.M. Lawan

Abstract Accurate and reliable wind speed and direction prediction is one of the necessary concepts in implementing a wind energy system. In this paper, meteorological and geographical variables were modeled via artificial neural networks (ANNs), taking terrain elevation and roughness class into account. The feedforward neural network (FFNN) with back propagation trained with Levenberg–Marquardt algorithm was utilized, with wind speed and direction as the target function in each model. The results obtained using the formulated topographical models showed a regression value R in the range of 0.8256–0.9883. The optimum network based on the lower mean square error and fast computation time was 9-152-1. Thus, the developed topographical feedforward neural network (T-FFNN) is efficient to predict the wind speed and direction properly.

Keywords Wind speed · Wind direction · Neural network · Sarawak

1 Introduction

Renewable energy resources are the major competitor of the fossil fuels such as coal, gas, and petroleum. Fossil fuel depletes with time, moreover, the resources are available in some regions around the world. Wind power is an indirect solar potential, which is clean, freely available, environmentally friendly, widely distributed, and naturally abundant almost anywhere around the globe. It has been applied decades

S.M. Lawan (✉) · W.A.W.Z. Abidin
Department of Electrical and Electronic Engineering, Faculty of Engineering, Universiti Malaysia Sarawak (UNIMAS), Sarawak, Malaysia
e-mail: 13010004@siswa.unimas.my

S.M. Lawan
Department of Electrical Engineering, Kano University of Science and Technology, Wudil, Nigeria

S. Lawan · A.M. Lawan
Faculty of Science, Department of Mathematical Science, Bayero University Kano (BUK), P.M.B 3011 Kano, Nigeria

A. Kılıçman et al. (eds.), *Recent Advances in Mathematical Sciences*,
DOI 10.1007/978-981-10-0519-0_7

ago for sailing ships, machine grinding, windmills, and crop handling. Recently, it has become popular for electrical power generation.

The development of wind energy has reached a large scale in terms of annual installed capacity. Large-scale wind farms are linked to the electrical power transmission lines; meanwhile, small wind turbines rated from few watts up to 10 kW are being used for stand-alone application to provide electricity to the isolated, remote, and rural locations [1–3]. Notwithstanding, many professionals have claimed that the wind potential remains unharnessed. In fact, 2014 is another remarkable year for wind turbine installation. Currently, wind turbine producers are yet to meet up with the present demands [4, 5].

Wind resource assessment (WRA), micrositing and sizing of wind turbines are prerequisite requirements that must be performed during the technical feasibility stage. The essential aspect that needs to be considered is the stochastic and unpredictable nature of wind speed. It is well known that a small deviation of wind speed will lead to a large error in the wind power output [6]. Because of this, wind speed prediction is essential for analyzing the performance of wind turbine system.

Many published studies have demonstrated the usefulness of wind speed and direction prediction models and this helps in siting of wind energy systems. Those techniques are usually classified into persistence, numeric weather prediction (NWP), statistical, stochastic, and method based on soft computing (artificial neural network (ANN) and fuzzy logic).

In the persistence approach, the accuracy drops with an increase in prediction time horizons. NWP method was based on the kinematics equations, which involve meteorological parameters. Examples, AIOLOS mass-consistent code numerical models [7]. COMPLEX and NOABL models were tested in the united kingdom (UK) [8]. Modified AIOLOS based on thermal stratification [9]. Multigrid solution via three-dimensional solver [10]. Moreover, according to [7], on the account of unpredictable nature of wind speed, it is difficult to generate a reliable algorithm that will take into account of all those irregularities, with acceptable accuracy. Furthermore, it has found that no mathematical model either physical or numeric will give a perfect definitive solution [11]. Meteorologist ordinarily uses the method to predict future samples.

A statistical and stochastic strategy mainly aims short samples. In the era of soft computing, neural network (NN), support vector machine (SVM), simulated annealing (SA) and fuzzy logic are found to be more appropriate [12–16].

On the other hand, in order to guarantee the prediction reliability for nonlinear time series, these techniques have some difficulties such as the problems in selecting input parameters, the complexity of computation time, and so on. In particular, one of the issues of fuzzy logic is the difficulty to determine the precise weights and the degree of each rule; moreover, it involves fuzzy sets and interval numbers. The application of ANN presents a series of advantages such as self-adaptive, parallelism, simple structure, rapid training speed, fast convergence fault tolerance. Based on these advantages, the ANN model has been widely formulated for various nonlinear applications [17]. The application of NN for the prediction of wind speed varies depending on the prediction timescale, that is, from short term up to long term. Many researchers [4, 18–21] have applied different forms of NN for wind speed prediction using different meteorological and geographical data. However, none of the listed studies have considered the influence of terrain shape and roughness changes. Nowadays, special attention has been put into the prediction of nonlinear and nonstationary time-varying data using ANNs, because of their widespread approximation functionality. It is generally acknowledged that ANN is the most suitable for the wind speed prediction [2, 22, 23]. The main objective of this paper is to present a novel ANN models named topographical feedforward neural network (T-FFNN) for predicting monthly wind speed and direction for Sarawak is proposed.

This research was carried out based upon the obtainable average hourly data at eight ground stations under the control of the Malaysia Meteorological Department (MMD). The rest of the paper is structured as follows. First, in depth wind speed and direction prediction using an extended proposed ANN and geographical information system (GIS) assisted methodology are discussed. Subsequently, the predicted and developed wind map results are analyzed. Lastly, purposeful findings are drawn.

2 Wind Speed and Direction Prediction Using ANN

Wide application of wind energy system necessitates accurate and precise estimation of wind speed/direction, which has a direct effect on the energy output. Traditionally, this can only be possible by using anemometer and wind vane to carry out experiments. However, in rural and remote areas, where obstacle-free wind resources are available, these devices or wind monitoring stations are not available. The goal of this paper is to provide a solution for estimating the abundant resource based on the knowledge of existing wind station in Sarawak, Malaysia. In fact, prediction tool is one of the best approaches for estimating wind speed and direction in non-monitored locations, which can help in designing of the wind energy systems, prior to microsizing and siting.

In wind engineering, ANNs have been widely used for the prediction of wind speed where a wind turbine is expected to operate [2, 22, 23].

ANNs are mathematical techniques constructed to perform a variety task, such as incremental learning, pattern recognition, process control data mining prediction, and financial modeling. They learn by examples and produce a future unseen data. ANN mimics human being brain system, which consist of layers of parallel element unit called neurons.

The neurons are connected to large number of weights over which the signal is passing; the neuron receives the input through the input layer and multiple by a weight generally performs nonlinear operations and produce an output. The most successful ANN used for the prediction is feedforward neural network (FFNN) using log-sigmoid [14, 15]. For these reasons, one hidden layer FFNN with backpropagation was adopted in this paper to examine the complex dependency of wind speed in the locations where wind station is not available, taking terrain shape and roughness height into consideration.

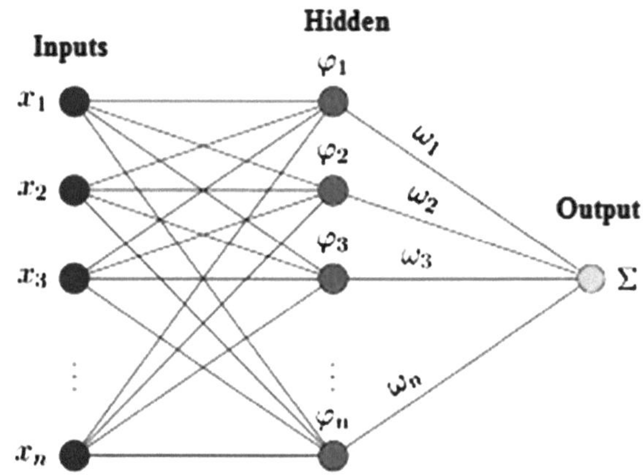

Fig. 1 T-FFNN for wind speed/direction prediction

Theoretically, it has been validated that a single layer design is satisfactory to approximate any nonlinear problems [5, 24]. Figure 1 shows a fully connected three layers topology of the proposed model for wind speed and direction prediction. Fully connected implies that the output from each of the input and hidden layer is distributed to all of the neurons in the subsequent layer. However, feedforward signifies that the network has not any directed cycle.

Although, there is no restriction regarding the adoption of transfer functions, it can be any mathematical function such as tangent, hyperbolic, logarithm, or combination of both like log-sigmoid, hyperbolic tangent, or Purelin. In this paper, log-sigmoid and Purelin (Fig. 2, [2]), activation functions were selected in order to obtain the differential function between the output and input variables, whose mathematical formulas are expressed in Eqs. 1 and 2 accordingly [25].

$$f(x) = \frac{1}{1 + e^{-x}} \tag{1}$$

$$f(x) = x \tag{2}$$

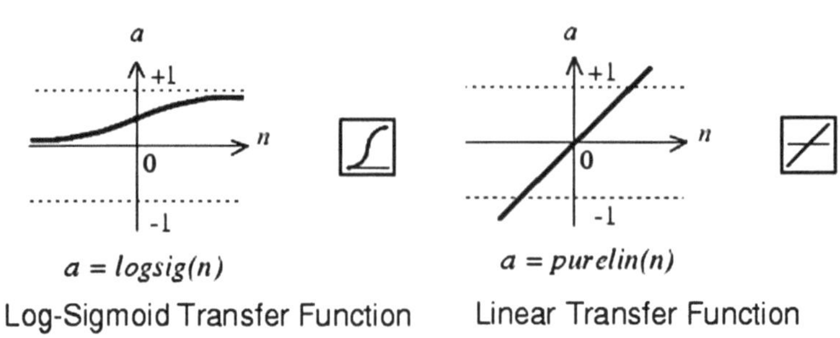

Fig. 2 Log-Sigmoid and purelin activation functions

Yet another most important phase in the course of designing an ANN is the training phase, simply because ANN is trained to solve a problem instead of programmed to accomplish this. The supervised training was adopted in this study. Monthly average data were feed in and the network learned by comparing the measured with the estimated. The error difference is propagated back from the output layer via hidden layer to the input layer and the weights on the connection between the neurons are updated as the error is backpropagated using Levenberg–Marquardt (LM). The goal of selecting LM was to assure speedy processing and furthermore to overwhelm the slow convergence associated with the conventional algorithm such as descent gradients and resilient propagation.

FFNN can mathematically approximate multivariate function to any level of accuracy if an adequate number of the hidden layer neurons are available. To avoid underfitting and overfitting in determining the number of optimal neurons in the hidden layer. This research work considered that the number of weights must not exceed the data used for the training. Hence, 9-7-1 topology was used with a step of five in all the designed models.

A MATLAB editor was used to the write the scripts file and implemented in the NN Toolbox. The developed models have 9 inputs latitude, longitude, altitude, and month of the year, temperature, atmospheric pressure, temperature and relative humidity, terrain shape and roughness height. Meanwhile, the output is the monthly wind speed for the wind speed prediction. In the case of wind direction, as the objection function (model 2). The network has latitude, longitude, altitude and month of the year, temperature, atmospheric pressure, temperature, and wind speed, terrain shape and roughness height. After the ANN training, the network was simulated to get the weights /biases, and can be applied to formulate the mathematical function.

The appropriateness of the designed models was assessed according to two statistical methods, the correlation coefficient (*R*) and mean absolute percentage error (*MAPE*). These values are mathematically described by the following equation [18, 26].

$$R = \frac{\sum_{i=1}^{N} \left(t_i - \bar{t}\right).\left(o_i - \bar{o}\right)}{\sqrt{\sum_{i=1}^{N} \left(t_i - \bar{t}\right)^2.\sqrt{\sum_{i=1}^{N} \left(o_i - \bar{o}\right)^2}}} \tag{3}$$

$$MAPE(\%) = \left(\frac{1}{N}\sum_{i=1}^{N}\left|\frac{t_i - o_i}{t_i}\right|\right)100 \tag{4}$$

where N is the number of data, and t_i, o_i are target value and ANN predicted value, respectively, of one data point i. The bars indicate the average value.

3 Isovent Lines Mapping and Maps Production

Wind energy mapping of the study area was generated using ARCGIS 9.3 software. The base maps of Sarawak at 1: 1,25,000 were digitized in order to obtain the shape files. The coordinate of the ground and predicted station were converted from degrees, minutes, seconds to the decimal unit. The World Geodetic System (WGS) of 1984 was applied for the definition of the coordination system and it was used in generating contour lines. Series of interpolation were conducted to select the best method that fits the wind speed and energy data of the studied area. It was found that Kriging method provides better accuracy for the spatial analysis carried out. This methodology has been verified in various studies [27, 28].

4 Results and Discussion

The employed data consist of 3650 daily records from Kuching, Miri, Sibu, Bintulu, and Sri Aman for the period of ten years (2003–2012). In addition, for the remaining stations, Kapit, Limbang and Mulu, the period of observation is 5 years (2008–2012). For the first five stations, the data were segmented from 2003–2009, 2010–2011, and 2012. For the last three stations, 2008–2009, 2010–2011, and 2012 for the training, testing, and validation. Prior to the training, all the data employed were scaled to the range of [−1, 1].

4.1 Topographical Simulation Models

Forty networks were designed and trained. The training was performed according to the mean squared error (MSE). It was identified that the optimum network in terms of fast convergence and lower MSE was 9-152-1 and 9-94-1 for the wind speed and direction models. The number of epochs was varied from 0 to 1000 in a step of one. It was realized that no further improvement could be made once the training reaches between 996 and 1000 epochs. Hence, the training was stopped. Figure 3 shows the error model function obtained during the training of one station. It can be noted from the figure that the after 1000 epochs the MSE reduces drastically between the ANN and measured data. The realized MSE was 0.043821. It should be observed that extended training would result in the ANN to remember training data, which ends up in poor generalization capability of the ANN model.

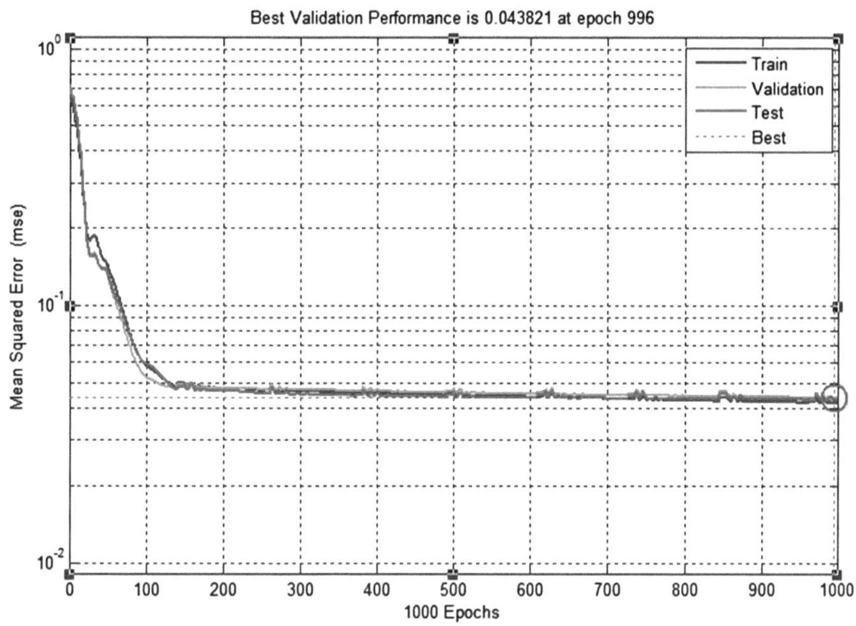

Fig. 3 Sample of the error function during the training (color figure online)

Pertaining to the data normalization, activation function used in the hidden layer and by using the weights and biases generated after the training, the equation for calculating wind speed and direction in the case of one station becomes:

$$v_m = 1.098\left[\left\{\left(\frac{2.907}{1+e^{-2.907(XW_1+B_1)}}\right)W_2 + B_2\right\} + 1\right] + 0.337 \tag{5}$$

$$w_D = 0.035\left[\left\{\left(\frac{0.072}{1+e^{-0.062(XW_1+B_1)}}\right)W_2 + B_2\right\} + 1\right] + 10.789 \tag{6}$$

where V_m, W_D: are the monthly wind speed/direction, W_1: Weight between the input and hidden layer, W_2: Weight between the hidden layer and output layer, B_1: Biases of the hidden layer, B_1: Biases in the output layer B_2, X: is the column vector, which contains normalized values of nine input variables.

In accordance with the results displayed in Table 1. The suggested models have the satisfactory precision for calculating wind speed and direction. It is interested to observe that, the T-FFNN models showed improved outcomes on the test data, which verifies the high generalization of the development process. For this reason,

Table 1 T-FFNN prediction accuaracy

Model	Data for the development		Data for model test	
	R	MAPE	R	MAPE
T-FFNN-1	0.8256–0.9871	5.46	0.9126–0.9671	4.69
T-FFNN-2	0.90123–0.985	4.29	0.9432–0.9883	3.87

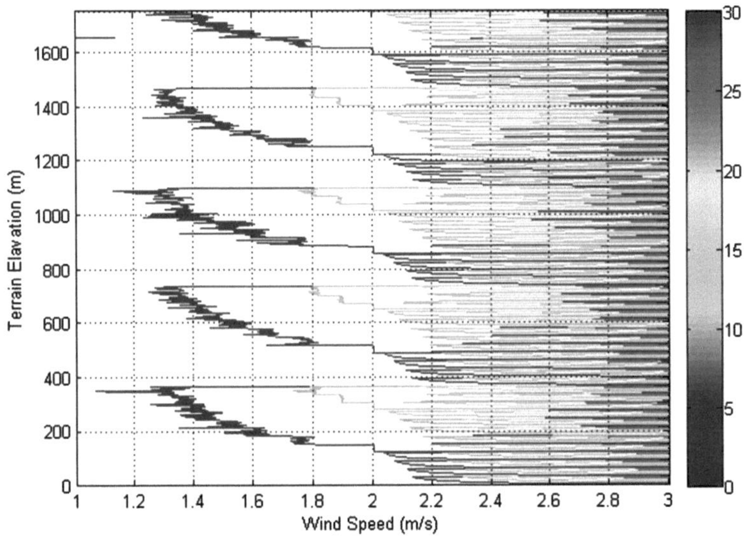

Fig. 4 Interaction effect between wind speed and terrain elevation at $v = 1\!-\!3$ m/s

the formulated models can be utilized efficiently to examine the relation between geographical, topographical, and meteorological variables on the wind speed and direction.

To this aim, Eq. 5 was plotted to generate a contours map response of wind speed for change in terrain condition (the shape of the terrain and roughness class) see Fig. 4. It is clear that, the terrain varies depending on the location, and wind speed is affected by the terrain and changes with the roughness of the terrain such as forest, lake, city, etc. From the figure, it is noticeable that the changes are more pronounce at higher wind speed, however, wind speed in the range of 1 m/s to about 2.2 m/s (class from 0–5 and 5–10) has limited effect on the wind flow within the studied area.

4.2 Wind Atlas Map of Sarawak at Various Elevations

The isovents map of Sarawak was developed based on the procedures discussed in the previous section. The generated map was based on the long-term measured and

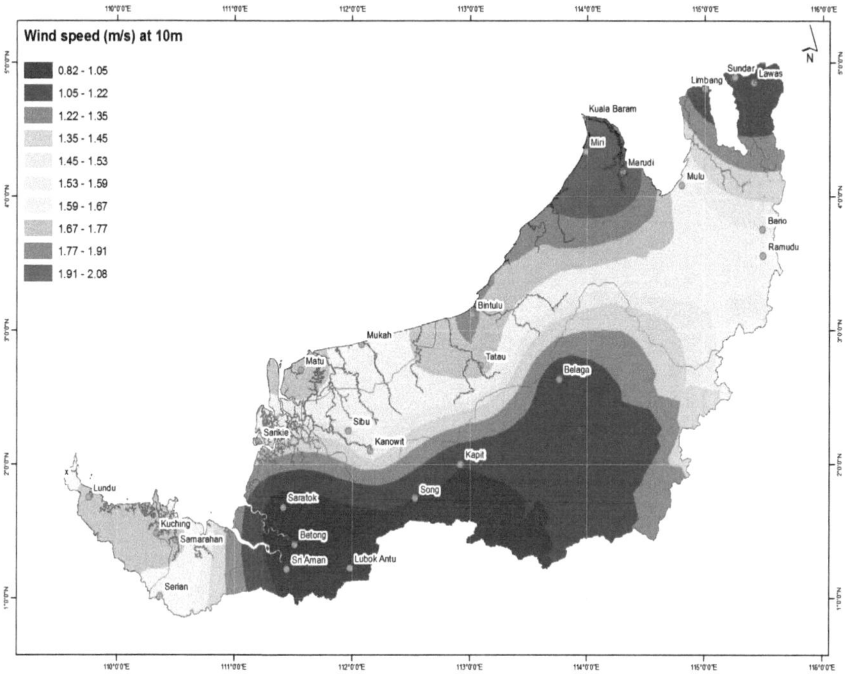

Fig. 5 Long-term wind speed map of Sarawak (color figure online)

predicted wind speed at 10–40 m in heights. Sample of the wind speed, power, and energy density maps are depicted in Figs. 5, 6 and 7 at 10 m in elevation. In all the listed figures, the areas marked red and blue represent highest and lowest potentials respectively.

5 Conclusion

In this paper, the idea of a new wind speed and direction prediction modeling based on the NN considering terrain shape is introduced. The proposed T-FFNN has been analyzed mathematically. The prediction models were used in conjunction with ground-based wind station to construct an isovents wind map of Sarawak that will be useful to policymakers, wind turbine manufacturers, potential investor and structural and bridge engineers.

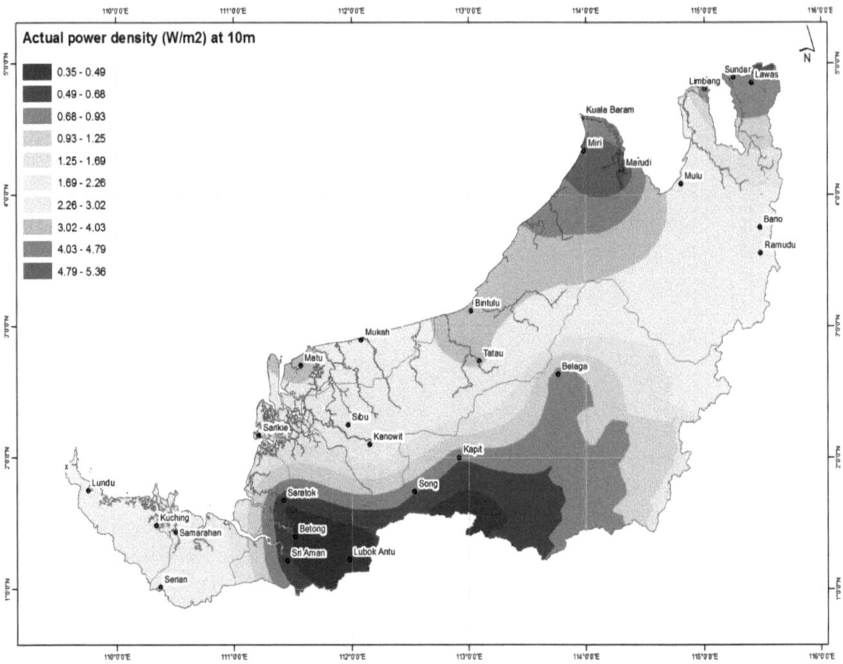

Fig. 6 Long-term power density map of Sarawak (color figure online)

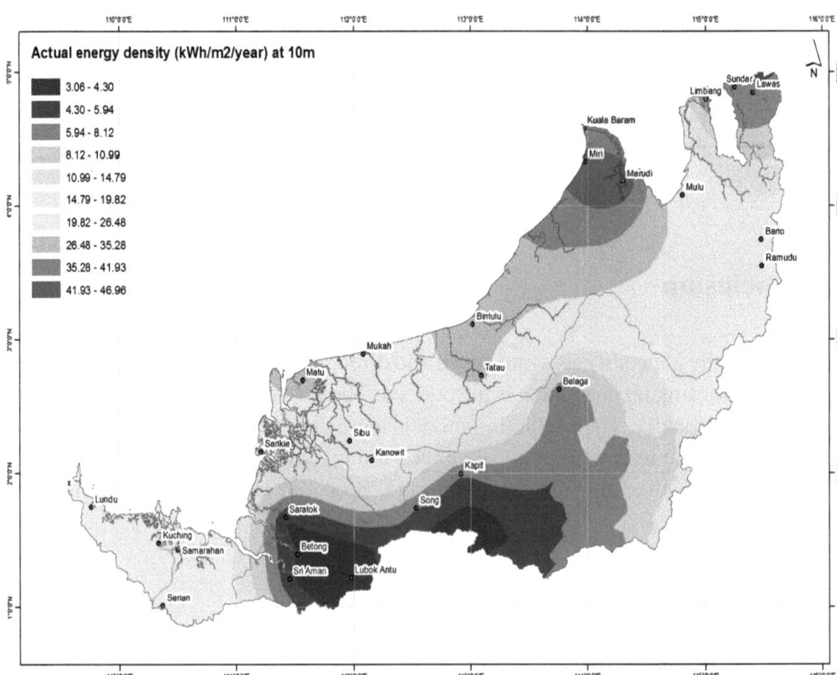

Fig. 7 Long-term energy density map of Sarawak (color figure online)

Acknowledgments The authors duly thanked the support of Universiti Malaysia Sarawak (UNIMAS) who has supported the data used in the research.

References

1. Mabel, C.M., Fernandez, E.: Analysis of wind power generation and prediction using ANN: a case study. Renew. Energy **33**, 986–992 (2008)
2. Muhammad, S.L., Abidin, W.A.W.Z., Chai, W.Y., Baharun, A., Masri, T.: Development of wind mapping based on artificial neural network (ANN) for energy exploration in Sarawak. Int. J. Renew. Energy Res. (IJRER) **4**, 618–627 (2014)
3. Azad, A.K., Rasul, M.G., Yusaf, T.: Statistical diagnosis of the best weibull methods for wind power assessment for agricultural applications. Energies **7**, 3056–3085 (2014)
4. Abbes, M., Belhadj, J.: Development of a methodology for wind energy estimation and wind park design. J. Renew. Sustain. Energy **6**, 053103 (2014)
5. Anvari-Moghaddam, A., Monsef, H., Rahimi-Kian, A., Nance, H.: Feasibility study of a novel methodology for solar radiation prediction on an hourly time scale: A case study in Plymouth, United Kingdom. J. Renew. Sustain. Energy **6**, 033107 (2014)
6. Lawan, S., Abidin, W., Chai, W., Baharun, A., Masri, T.: The status of wind resource assessment (WRA) techniques, wind energy potential and utilisation in Malaysia and other countries. J. Eng. Appl. Sci. **8**, (2013)
7. Ratto, C., Festa, R., Romeo, C., Frumento, O., Galluzzi, M.: Mass-consistent models for wind fields over complex terrain: the state of the art. Environ. Softw. **9**, 247–268 (1994)
8. Guo, X., Palutikof, J.: A study of two mass-consistent models: problems and possible solutions. Bound. Layer Meteorol. **53**, 303–332 (1990)
9. Focken, U., Heinemann, D., Waldl, H.P.: Wind assessment in complex terrain with the numeric model Aiolos: implementation of the influence of roughness changes and stability. In: EWEC-Conference, pp. 1173–1176 (1999)
10. Dinar, N.: Mass consistent models for wind distribution in complex terrain–Fast algorithms for three dimensional problems. In: Boundary Layer Structure, ed, pp. 177–199, Springer, Bosdon (1984)
11. Focken, U., Lange, M.: Physical approach to short-term wind power prediction. Springer, New York (2006)
12. Ahmad, A., Anderson, T.: Global Solar Radiation Prediction Using Artificial Neural Network Models for New Zealand (2014)
13. Ak, R., Li, Y., Vitelli, V., Zio, E.: Estimation of wind speed prediction intervals by multi-objective genetic algorithms and neural networks. In: Acts of the XLVI Scientific Meeting of the Italian Statistical Society, Rome, Italy (2012)
14. Alkhatib, A, Heire. S., Kurt M.: Detailed analysis for implementing a short term wind speed prediction tool using artificial neural networks. In: International Journal on Advances in Networks and Services, vol. 5, pp. 149–158 (2012)
15. Anand, A.P., Saravanan, R., Muthaiah, R.: Threshold prediction of a cyclostationary feature detection process using an artificial neural network. In: International Journal of Engineering & Technology, vol. 5, pp. 0975–4024 (2013)
16. Kalogirou, S.A.: Artificial neural networksin renewable energy systems applications: A review. Renew. Sustain. Energy Rev. **5**, 373–401 (2001)
17. Kazemi, K., Moradi, S., Asoodeh, M.: A neural network based model for prediction of saturation pressure from molecular components of crude oil. Energy sources, Part A: Recovery, utilization, and environmental effects **35**, 1039–1045 (2013)
18. Khatib, T., Alsadi, S.: Modeling of wind speed for palestine using artificial neural network. J. Appl. Sci. **11**, 2634–2639 (2011)

19. Fadare, D.: The application of artificial neural networks to mapping of wind speed profile for energy application in Nigeria. Appl. Energy **87**, 934–942 (2010)
20. Sözen, A., Arcaklioglu, E., Özalp, M., Kanit, E.G.: Use of artificial neural networks for mapping of solar potential in Turkey. Appl. Energy **77**, 273–286 (2004)
21. Ozgonenel, O., Thomas, D.W.: Short-term wind speed estimation based on weather data. Turkish J. Elect. Eng. Comput. Sci. **20**, 335–346 (2012)
22. Lin, W.-M., Hong, C.-M.: A new elman neural network-based control algorithm for adjustable-pitch variable-speed wind-energy conversion systems. IEEE Trans. Power Electron. **26**, 473–481 (2011)
23. Musyafa, A., Cholifah, B., Dharma, A., Robandi, I.: Local short term wind speed prediction in the region Nganjuk City (East Java) using neural network. In: Local Short Term Wind Speed Prediction in the Region Nganjuk City (East Java) Using Neural Network (2013)
24. Philippopoulos, K., Deligiorgi, D., Kouroupetroglou, G.: Artificial neural network modeling of relative humidity and air temperature spatial and temporal distributions over complex terrains. In: Pattern Recognition Applications and Methods, ed, pp. 171–187, Springer, New York (2015)
25. Mohandes, M., Balghonaim, A., Kassas, M., Rehman, S., Halawani, T.: Use of radial basis functions for estimating monthly mean daily solar radiation. Solar Energy **68**, 161–168 (2000)
26. Madić, M., Radovanović, M.: An artificial intelligence approach for the prediction of surface roughness in CO_2 laser cutting. J. Eng. Sci. Technol. **7**, 679–689 (2012)
27. Ramachandra, T., Shruthi, B.: Wind energy potential mapping in Karnataka, India, Using GIS. Energy Convers. Manag. **46**, 1561–1578 (2005)
28. Bui, T.Q., Nguyen, T.N., Nguyen-Dang, H.: A moving kriging interpolation-based meshless method for numerical simulation of Kirchhoff plate problems. Int. J. Num. Methods Eng. **77**, 1371–1395 (2009)

Stability Analysis of Dengue Disease Using Host–Vector Model

Eminugroho Ratna Sari

Abstract The previous model of dengue disease was discussed only in host population using a simple SIR model. In fact, vectors provide an important role in the spread of dengue fever since a host can be infected by the vector. Hence it is reasonable to build a model of dengue disease in host and vector population. From the model, there are two kinds of equilibrium point: disease free and endemic. Solution behavior of model can be analyzed using the changes of basic reproduction number which is obtained by next generating matrix. If basic reproduction number is less or equal than one, then using LaSalle–Lyapunov Theorem, it is shown that the disease-free equilibrium is globally asymptotically stable. If the basic reproductive number is greater than one, then using Routh–Hurwitz Condition, it is shown that the endemic equilibrium is locally asymptotically stable. In the end, we present the numerical solution with MAPLE.

Keywords Dengue · Host–vector · Basic reproduction · Stability

1 Introduction

Dengue fever is caused by a virus and *Aedes aegypti* mosquito is the responsible vector to transmit it. Dengue virus has four different serotypes namely DEN1, DEN2, DEN3, and DEN4. It is possible to become infected multiple times by different serotype of virus. The virus has an incubation period of 4–10 days in the susceptible vector. After the incubation period, the mosquito is called infected vector. If the infected vector bites susceptible human, then the virus exists in the human body until it is ready to transmit to the mosquito. Infected humans can transmit the infection for 4–5 days, through *Aedes aegypti* after their first symptoms appear [1].

Dengue fever is a common disease that occurs in many tropical and subtropical countries. In the 1780s, the first recognized epidemic of dengue occurred almost

E.R. Sari (✉)
Mathematics Education Department, Faculty of Mathematics and Natural Science, Yogyakarta State University, Karangmalang, Yogyakarta 55281, Indonesia
e-mail: eminugrohosari@gmail.com

A. Kılıçman et al. (eds.), *Recent Advances in Mathematical Sciences*,
DOI 10.1007/978-981-10-0519-0_8

simultaneously in Asia, Africa, and North America, and nowadays is a disease widely found [2]. According to [1], every year 50–100 million people are affected with dengue fever and the most cases occur in Asia. Meanwhile, during 1968–2009, WHO noted that in Southeast Asia, Indonesia has the highest dengue cases. In Indonesia, dengue fever was first discovered in Surabaya in 1968, where as many as 58 people were infected and 24 died. It means the mortality rate was 41.3 %, and after that the disease spread throughout the country [3].

Mathematical modeling of dengue transmission contributes greatly to predict and control the disease spread. Modeling of two types of viruses has been discussed in [4] and the time needed for an initial condition to reach a certain number of infective has been estimated in [5]. The transmission model with vaccination has been identified in [2, 6, 7]. Furthermore, SIR model to analyze dengue behavior focuses in host population has used in [8]. In this paper, first, we discuss the dynamics of dengue disease focuses in one type of virus that refers to study of Esteva [9] with detailed derivation of a model and its dynamical analysis. Then, we modified the first model by adding the assumption that the host population can lose its immunity, the death rate caused by the disease, and there is the percentage of infected humans but incapable of transmitting the disease.

2 Model Formulation Focuses in One Type of Virus

Let N_H be the human (host) population size and N_V be the vector population size. In this model, host population is divided into three classes, i.e., the number of susceptible, infectious, and recovered at time t. It is denoted by $S_H(t)$, $I_H(t)$, $R_H(t)$, respectively. In this section, we only focus on a single type of virus, so for every infected human will has permanent immunity to it. Because the infection period of mosquitoes will end by its death, it never recovers from the infection [10]. Hence, we divide the vector population into two classes $S_V(t)$ and $I_V(t)$ which denote the number of susceptible and infectious vectors at time t. Furthermore, we define some parameters which are summarized in Table 1.

We assume an equal constant size for birth rate and natural death rate in the human population. Death due to disease is assumed to be negligible. We also assume that the human population will be immune after recovering from dengue disease, every birth in human population enters the susceptible class, the infection rate of human population which according to bite of vector $\left(\frac{\beta_H b}{N_H}\right)$ is proportional to susceptible human and infectious vector. For the vector population, we denote A as a constant recruitment rate, the infection rates of vector population $\left(\frac{\beta_V b}{N_H}\right)$ are proportional to susceptible vector and infectious human.

From the assumptions, we obtain the diagram transfer as shown in Fig. 1.

Table 1 Parameters which are used in the model

Parameter	Biological meaning	Range of values	Reference
μ_H	Per capita birth/naturally death rate of human	0.0143–0.0167 per year	[11]
μ_V	Per capita death rate of vector	0.02–0.09 per day	[11]
b	The average number of bites per vector per day	0–1	[11]
β_H	Transmission probability from vector to human	0.75	[11]
β_V	Transmission probability from human to vector	1	[9]
γ_H	Infected human get treated (or recover) rate	0.083–0.25 per day	[11]

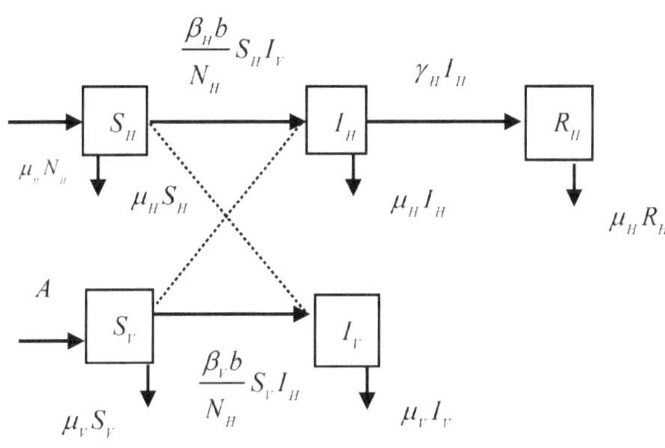

Fig. 1 Diagram transfer for dengue disease model

Using Fig. 1, we have the mathematical model of dengue disease in ODEs system as follows:

$$
\begin{aligned}
\frac{dS_H(t)}{dt} &= \mu_H N_H - \frac{\beta_H b}{N_H} S_H I_V - \mu_H S_H \\
\frac{dI_H(t)}{dt} &= \frac{\beta_H b}{N_H} S_H I_V - (\mu_H + \gamma_H) I_H \\
\frac{dR_H(t)}{dt} &= \gamma_H I_H - \mu_H R_H \\
\frac{dS_V(t)}{dt} &= A - \frac{\beta_V b}{N_H} S_V I_H - \mu_V S_V \\
\frac{dI_V(t)}{dt} &= \frac{\beta_V b}{N_H} S_V I_H - \mu_V I_V
\end{aligned}
\tag{1}
$$

with $S_H + I_H + R_H = N_H$ and $S_V + I_V = N_V$. Since μ_V denotes the per capita death rate of the vector, $\mu_V N_V$ denotes the total deaths in the vector population.

Furthermore, the true change rate of vector population is denoted by $\frac{dN_V}{dt}$, that is the recruitment rate minus the total death of vector, $\frac{dN_V}{dt} = A - \mu_V N_V$. As $t \to \infty$, the solution N_V approaches $\frac{A}{\mu_V}$. Then we can define

$$S_H + I_H + R_H = N_H \text{ and } S_V + I_V = \frac{A}{\mu_V}.$$

Therefore, we can discuss the behavior of solution of system (1) in this set. Furthermore, we can introduce the proportion

$$s_H = \frac{S_H}{N_H}, \quad i_H = \frac{I_H}{N_H}, \quad r_H = \frac{R_H}{N_H}, \quad s_V = \frac{S_V}{A/\mu_V}, \quad i_V = \frac{I_V}{A/\mu_V}.$$

We also use the relation $r_H = 1 - (s_H + i_H)$ and $s_V = 1 - i_V$. Then the first equation of system (1) can be written as

$$\begin{aligned} \frac{1}{N_H}\left(\frac{dS_H(t)}{dt}\right) &= \frac{1}{N_H}\left(\mu_H N_H - \frac{\beta_H b}{N_H} S_H I_V - \mu_H S_H\right) \\ \Leftrightarrow \frac{ds_H(t)}{dt} &= \mu_H - \frac{\beta_H b}{N_H} s_H (A/\mu_V)\, i_V - \mu_H s_H \\ \Leftrightarrow \frac{ds_H(t)}{dt} &= \mu_H (1 - s_H) - \frac{\beta_H b (A/\mu_V)}{N_H} s_H i_V. \end{aligned} \tag{2a}$$

Similarly, the second equation of system (1) can be written as

$$\frac{di_H(t)}{dt} = \frac{\beta_H b (A/\mu_V)}{N_H} s_H i_V - (\mu_H + \gamma_H)\, i_H. \tag{2b}$$

For the last equation of system (1), we use the proportion and relation as introduced before to obtain

$$\frac{di_V}{dt} = \beta_V b s_V i_H - \mu_V i_V = \beta_V b (1 - i_V)\, i_H - \mu_V i_V. \tag{2c}$$

In the next section, we provide the analysis of system (2).

2.1 Determination of Equilibrium Point

The equilibrium points of system (2) will be explained in the Lemma 1.

Lemma 1 *(a) If $i_H = 0$, then System (2) has a disease-free equilibrium point, $E_0 =$ (1, 0, 0).*

(b) *If* $i_H \neq 0$*, then System (2) has an endemic equilibrium point,* $E_1 = \left(\widehat{s}_H, \widehat{i}_H, \widehat{i}_V\right)$*, where*

$$\widehat{s}_H = \frac{\frac{b\beta_V}{\mu_V} + \frac{\gamma_H+\mu_H}{\mu_H}}{\frac{b\beta_V}{\mu_V} + \left(\frac{\gamma_H+\mu_H}{\mu_H}\right)\widehat{R}_0},\ \widehat{i}_H = \frac{\widehat{R}_0 - 1}{\left(\frac{b\beta_V}{\mu_V} + \frac{\gamma_H+\mu_H}{\mu_H}\widehat{R}_0\right)},\ \widehat{i}_V = \frac{\left(\frac{\beta_V b}{\mu_V}\right)\left(\widehat{R}_0 - 1\right)}{\left(\frac{\beta_V b}{\mu_V} + \frac{\gamma_H+\mu_H}{\mu_H}\right)\widehat{R}_0}.$$

Proof With the right-hand side equal to zero for (2c) can be obtained

$$i_V = \frac{\beta_V b i_H}{(\beta_V b i_H + \mu_V)} = \frac{\frac{\beta_V b}{\mu_V} i_H}{\frac{\beta_V b}{\mu_V} i_H + 1}. \tag{3}$$

From (2a) with the right-hand side equal to zero and substitute (3), then

$$s_H = \frac{\mu_H}{\frac{\beta_H b(A/\mu_V)}{N_H} i_V + \mu_H} = \frac{b\beta_V i_H + \mu_V}{b\beta_V i_H + \mu_V + \frac{b^2\beta_V\beta_H A/\mu_V i_H}{N_H\mu_H}}. \tag{4}$$

If $\widehat{R}_0 = \frac{\beta_H b(A/\mu_V)}{N_H(\mu_H+\gamma_H)}\frac{\beta_V b}{\mu_V}$, then Eq. 4 becomes

$$s_H = \frac{\left(\frac{b\beta_V}{\mu_V} i_H + 1\right)}{1 + \left(\frac{b\beta_V}{\mu_V} + \frac{\gamma_H+\mu_H}{\mu_H}\widehat{R}_0\right) i_H}. \tag{5}$$

Substituting Eqs. 3 and 5 in (2b) with the right-hand side equal to zero, we obtain

$$\frac{(\gamma_H + \mu_H)\widehat{R}_0\left(\frac{b\beta_V}{\mu_V} i_H^2 + i_H\right) i_H}{i_H\left(\frac{b\beta_V}{\mu_V} i_H + 1\right)\left(1 + \left(\frac{b\beta_V}{\mu_V} + \frac{\gamma_H+\mu_H}{\mu_H}\widehat{R}_0\right) i_H\right)} - (\mu_H + \gamma_H) i_H = 0$$

$$\Leftrightarrow -\left(\frac{b\beta_V}{\mu_V} + \frac{\gamma_H + \mu_H}{\mu_H}\widehat{R}_0\right) i_H^2 + \left(\widehat{R}_0 - 1\right) i_H = 0. \tag{6}$$

Solving Eq. 6, we obtain $i_H = 0$ or

$$i_H = \frac{\widehat{R}_0 - 1}{\left(\frac{b\beta_V}{\mu_V} + \frac{\gamma_H+\mu_H}{\mu_H}\widehat{R}_0\right)}. \tag{7}$$

There are two possibilities

(a) If $i_H = 0$, then from Eq. 3 $i_V = 0$. Moreover, from Eq. 4 it can be calculated that $s_H = 1$. Thus, the equilibrium point is $E_0 = (1, 0, 0)$, namely disease-free equilibrium since the number of infectious both human and vector tend to zero.

(b) If $i_H \neq 0$, in other words, the number of infectious humans exist, then we can obtain the number of infectious vector and the number of susceptible human also exist. If Eq. 7 is substituted in Eq. 3, then we obtain

$$i_V = \frac{\left(\frac{\beta_V b}{\mu_V}\right)\left(\widehat{R}_0 - 1\right)}{\left(\frac{\beta_V b}{\mu_V} + \frac{\gamma_H + \mu_H}{\mu_H}\right)\widehat{R}_0}. \tag{8}$$

If Eq. 7 is substituted in Eq. 5, then we obtain

$$s_H = \frac{\frac{b\beta_V}{\mu_V} + \frac{\gamma_H + \mu_H}{\mu_H}}{\frac{b\beta_V}{\mu_V} + \left(\frac{\gamma_H + \mu_H}{\mu_H}\right)\widehat{R}_0}. \tag{9}$$

The equilibrium point $E_1 = \left(\widehat{s}_H, \widehat{i}_H, \widehat{i}_V\right)$ with endemic equilibrium point $\widehat{s}_H, \widehat{i}_H, \widehat{i}_V$ are given in Eqs. 9, 7, 8 respectively. □

2.2 Basic Reproduction Number

Determination of the value of basic reproduction number (R_0) will be discussed in this section. Following the idea in [11–13], we use next-generation matrix to find the threshold value. The ability of the disease to invade population can be analyzed with this number.

First, we create matrix F as a matrix of gain terms of each class and evaluated at the disease-free equilibrium. Thus

$$F = \begin{bmatrix} \frac{\partial}{\partial i_H}\left(\frac{\beta_H b(A/\mu_H)}{N_H} s_H i_V\right) & \frac{\partial}{\partial i_H}\left(\beta_V b i_H\right) \\ \frac{\partial}{\partial i_V}\left(\frac{\beta_H b(A/\mu_H)}{N_H} s_H i_V\right) & \frac{\partial}{\partial i_V}\left(\beta_V b i_H\right) \end{bmatrix}_{\text{at }(1,0,0)}$$
$$= \begin{bmatrix} 0 & \beta_V b \\ \frac{\beta_H b(A/\mu_H)}{N_H} & 0 \end{bmatrix}$$

Then, we create matrix V as a matrix of loss terms of each class and evaluated at the disease-free equilibrium

$$V = \begin{bmatrix} \frac{\partial}{\partial i_H}\left(\mu_H + \gamma_H\right) i_H & \frac{\partial}{\partial i_H}\left(\mu_V + \beta_V b i_H\right) i_V \\ \frac{\partial}{\partial i_V}\left(\mu_H + \gamma_H\right) i_H & \frac{\partial}{\partial i_V}\left(\mu_V + \beta_V b i_H\right) i_V \end{bmatrix}_{\text{at }(1,0,0)}$$
$$= \begin{bmatrix} (\mu_H + \gamma_H) & 0 \\ 0 & (\mu_V) \end{bmatrix}$$

Therefore

$$V^{-1} = \frac{1}{(\mu_H + \gamma_H)\,\mu_V} \begin{bmatrix} \mu_V & 0 \\ 0 & \mu_H + \gamma_H \end{bmatrix} = \begin{bmatrix} \frac{1}{\mu_H+\gamma_H} & 0 \\ 0 & \frac{1}{\mu_V} \end{bmatrix}.$$

Matrix G is a product of F and V^{-1},

$$G = FV^{-1} = \begin{bmatrix} 0 & \frac{\beta_V b}{\mu_V} \\ \frac{\beta_H b(A/\mu_H)}{N_H(\mu_H+\gamma_H)} & 0 \end{bmatrix}.$$

The largest Eigen value from matrix G is $\sqrt{\frac{\beta_H b(A/\mu_H)}{N_H(\mu_H+\gamma_H)}\frac{\beta_V b}{\mu_V}}$. This is called basic reproduction number of System (2), denoted by R_0. If $\widehat{R}_0 = \frac{\beta_H b(A/\mu_H)}{N_H(\mu_H+\gamma_H)}\frac{\beta_V b}{\mu_V}$, then $R_0 = \sqrt{\widehat{R}_0}$. It is represented the average number of secondary cases that one case can produce if introduced into a susceptible population. The stability of equilibrium points can be explained in the next section.

2.3 *Detailed Explanation of Stability Analysis of Each Equilibrium Point*

Explanation about stability analysis will be given in Lemmas 2 and 3 as follows.

Lemma 2 *a. If $R_0 > 1$, then E_0 is unstable.*
b. If $R_0 \leq 1$, then E_0 is globally asymptotically stable.

Proof The Jacobian matrix at E_0 is

$$J(E_0) = \begin{bmatrix} -\mu_V & 0 & \frac{-b\beta_H A/\mu_V}{N_H} \\ 0 & -(\gamma_H + \mu_H) & \frac{b\beta_H A/\mu_V}{N_H} \\ 0 & b\beta_V & -\mu_V \end{bmatrix} \tag{10}$$

If λ denotes the eigenvalue and I denotes the identity matrix, then the eigenvalues of matrix (10) are determined by $|J(E_0) - \lambda I| = 0$. It implies

$$(-\mu_H - \lambda)\left(\lambda^2 + (\gamma_H + \mu_H + \mu_V)\lambda + \mu_V(\gamma_H + \mu_H) - b\beta_V\left(\frac{b\beta_H A/\mu_V}{N_H}\right)\right) = 0.$$

Then we obtain

$$\lambda = -\mu_H. \tag{11}$$

Also

$$\lambda^2 + (\gamma_H + \mu_H + \mu_V)\,\lambda + \mu_V\,(\gamma_H + \mu_H) - b\beta_V\left(\frac{b\beta_H A/\mu_V}{N_H}\right) = 0$$
$$\Leftrightarrow \lambda^2 + (\gamma_H + \mu_H + \mu_V)\,\lambda + \mu_V\,(\gamma_H + \mu_H)\left(1 - \widehat{R}_0\right) = 0$$

Solving the last equation, we get

$$\lambda = \frac{-(\gamma_H + \mu_H + \mu_V) \pm \sqrt{(\gamma_H + \mu_H + \mu_V)^2 - 4\mu_V\,(\gamma_H + \mu_H)\left(1 - \widehat{R}_0\right)}}{2}. \tag{12}$$

a. It implies if $\widehat{R}_0 > 1$, then one of the eigenvalue in (12), $\left(-(\gamma_H + \mu_H + \mu_V) + \sqrt{(\gamma_H + \mu_H + \mu_V)^2 - 4\mu_V\,(\gamma_H + \mu_H)\left(1 - \widehat{R}_0\right)}\right)\Big/2$, is positive. Stated that if $\widehat{R}_0 > 1$ means $R_0 > 1$, and the Lemma section a completes.
b. It is easy to show if $\widehat{R}_0 < 1$, then the eigenvalues in Eq. 12 are negative. Since the eigenvalues in Eqs. 11–12 are negative, E_0 is locally asymptotically stable. To explain the global stability, we use function $H : \mathbb{R}^3_{+,0} \to \mathbb{R}$, where $\mathbb{R}^3_{+,0} = \left\{(s_H, i_H, i_v) \in \mathbb{R}^3 : s_H, i_H, i_v \geq 0\right\}$ and

$$H\,(\mathbf{x}) = \frac{b\beta_H A/\mu_V}{\mu_V N_H} i_V + i_H. \tag{13}$$

Equation 13 has properties as follows:

1. Function H and its partial derivative are continue in $\mathbb{R}^3$.
2. Function H is positive definite i.e.
   a. For every $\mathbf{x} = (s_H, i_H, i_v) \in \mathbb{R}^3_{+,0}$, $H\,(\mathbf{x}) \geq 0$
   b. $H\,(\mathbf{x}) = 0 \Leftrightarrow \mathbf{x} = 0$
   c. If $\mathbf{x} \to \infty$, then $H\,(\mathbf{x}) \to \infty$
3. The derivative of H with respect to t is given by

$$\begin{aligned}\dot{H}\,(\mathbf{x}) &= \frac{\partial H}{\partial s_H}\frac{ds_H}{dt} + \frac{\partial H}{\partial i_H}\frac{di_H}{dt} + \frac{\partial H}{\partial i_V}\frac{di_V}{dt} \\ &= -\frac{b\beta_H A/\mu_V}{N_H}\,(1 - s_H)\,i_V - (\gamma_H + \mu_H)\left(1 - \widehat{R}_0\,(1 - i_V)\right) i_H\end{aligned}$$

   which less or equal to zero if $\widehat{R}_0 \leq 1$.
4. $\dot{H}\,(E_0) = 0$.

According to LaSalle–Lyapunov Theorem [14], the equilibrium point E_0 is globally asymptotically stable. Note that if $\widehat{R}_0 \leq 1$ it implies $R_0 \leq 1$, and the Lemma section b completes. □

Lemma 3 *If $R_0 > 1$, then the endemic equilibrium point, E_1, is locally asymptotically stable.*

Proof The Jacobian matrix at E_1 is

$$J(E_1) = \begin{bmatrix} \frac{-\mu_H\left(\frac{b\beta_V}{\mu_V}+\left(\frac{\gamma_H+\mu_H}{\mu_H}\right)\widehat{R}_0\right)}{\frac{b\beta_V}{\mu_V}+\frac{\gamma_H+\mu_H}{\mu_H}} & 0 & \frac{-(\gamma_H+\mu_H)\widehat{R}_0}{\frac{b\beta_V}{\mu_V}}\left(\frac{\frac{(\gamma_H+\mu_H)}{\mu_H}+\frac{b\beta_V}{\mu_V}}{\frac{(\gamma_H+\mu_H)\widehat{R}_0}{\mu_H}+\frac{b\beta_V}{\mu_V}}\right) \\ \frac{(\gamma_H+\mu_H)\left(\widehat{R}_0-1\right)}{\left(\frac{b\beta_V}{\mu_V}+\frac{\gamma_H+\mu_H}{\mu_H}\right)} & -(\gamma_H+\mu_H) & \frac{(\gamma_H+\mu_H)\widehat{R}_0}{\frac{b\beta_V}{\mu_V}}\left(\frac{\frac{(\gamma_H+\mu_H)}{\mu_H}+\frac{b\beta_V}{\mu_V}}{\frac{(\gamma_H+\mu_H)\widehat{R}_0}{\mu_H}+\frac{b\beta_V}{\mu_V}}\right) \\ 0 & \frac{b\beta_V}{\widehat{R}_0}\frac{\frac{(\gamma_H+\mu_H)\widehat{R}_0}{\mu_H}+\frac{b\beta_V}{\mu_V}}{\frac{(\gamma_H+\mu_H)}{\mu_H}+\frac{b\beta_V}{\mu_V}} & -\mu_V\widehat{R}_0\left(\frac{\frac{(\gamma_H+\mu_H)}{\mu_H}+\frac{b\beta_V}{\mu_V}}{\frac{(\gamma_H+\mu_H)\widehat{R}_0}{\mu_H}+\frac{b\beta_V}{\mu_V}}\right) \end{bmatrix} \tag{14}$$

If λ denotes the eigenvalue and I denotes the identity matrix, then the characteristic polynomials of matrix (14) are determined by $|J(E_1) - \lambda I|$. It implies

$$p(\lambda) = \lambda^3 + J_1\lambda^2 + J_2\lambda + J_3, \tag{15}$$

where

$$J_1 = \frac{\mu_H\left(\frac{b\beta_V}{\mu_V}+\frac{\gamma_H+\mu_H}{\mu_H}\widehat{R}_0\right)}{\frac{b\beta_V}{\mu_V}+\frac{\gamma_H+\mu_H}{\mu_H}} + (\gamma_H+\mu_H) + \mu_V\widehat{R}_0\left(\frac{\frac{b\beta_V}{\mu_V}+\frac{\gamma_H+\mu_H}{\mu_H}}{\frac{b\beta_V}{\mu_V}+\frac{\gamma_H+\mu_H}{\mu_H}\widehat{R}_0}\right),$$

$$J_2 = \mu_V(\gamma_H+\mu_H)\frac{b\beta_V}{\mu_V}\left(\frac{\widehat{R}_0-1}{\frac{b\beta_V}{\mu_V}+\frac{(\gamma_H+\mu_H)}{\mu_H}\widehat{R}_0}\right) + \mu_V\mu_H\widehat{R}_0$$
$$+ \mu_H(\gamma_H+\mu_H)\left(\frac{\frac{b\beta_V}{\mu_V}+\frac{(\gamma_H+\mu_H)}{\mu_H}\widehat{R}_0}{\frac{b\beta_V}{\mu_V}+\frac{(\gamma_H+\mu_H)}{\mu_H}}\right),$$

$$J_3 = \mu_V\mu_H(\gamma_H+\mu_H)\left(\widehat{R}_0-1\right).$$

Noted that if $\widehat{R}_0 > 1$, then J_1, J_2, J_3 are positive. Furthermore, it is easy to identify $J_1J_2 > J_3$, then by Routh–Hurwitz condition E_1 is locally asymptotically stable. □

Using parameter values presented in Table 1 and different initial points show that s_H decreases in a small period; meanwhile, both i_H and i_V increase significantly. It can be seen from Fig. 2 during 50 days to approach (0.09529, 0.00028, 0.00057) as

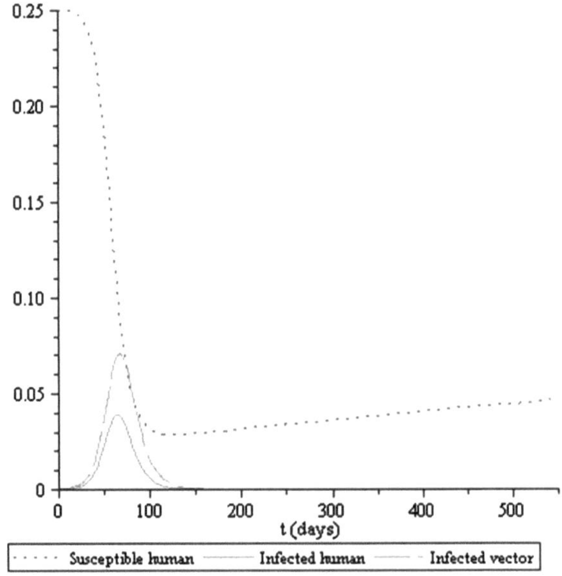

Fig. 2 Using $A = 5000$, $N_H = 10{,}000$, $s_H(0) = 0.25$, $i_H(0) = 0.0003$, $i_V(0) = 0.00005$

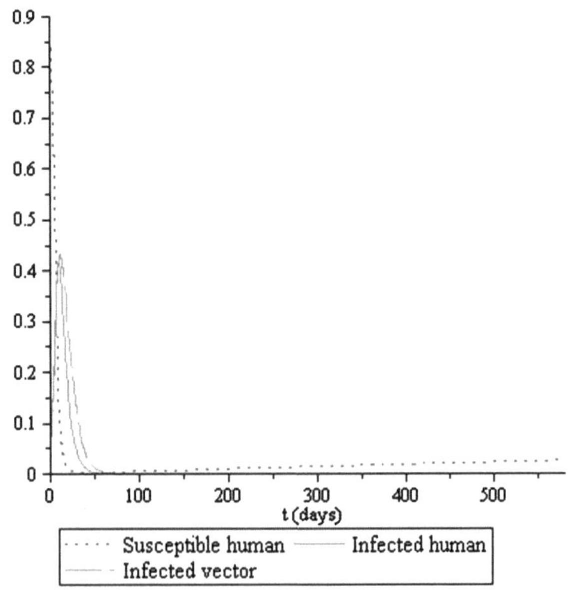

Fig. 3 Using $A = 5000$, $N_H = 10{,}000$, $s_H(0) = 0.9$, $i_H(0) = 0.05$, $i_V(0) = 0.05$

the endemic equilibrium point but we just need for about 30 days in Fig. 3 to make the solution oscillate around this point.

Figures 4 and 5 show that susceptible will increase to one and the infection proportion tends to zero. It shows us that the disease will die out.

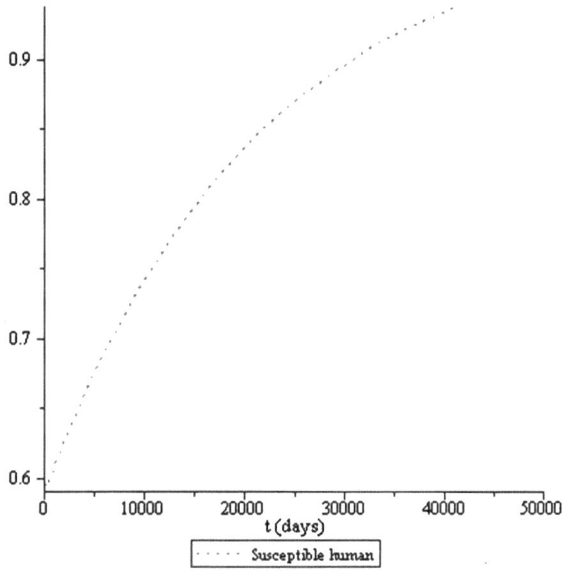

Fig. 4 Numerical solution of s_H using $A = 500$, $N_H = 10{,}000$

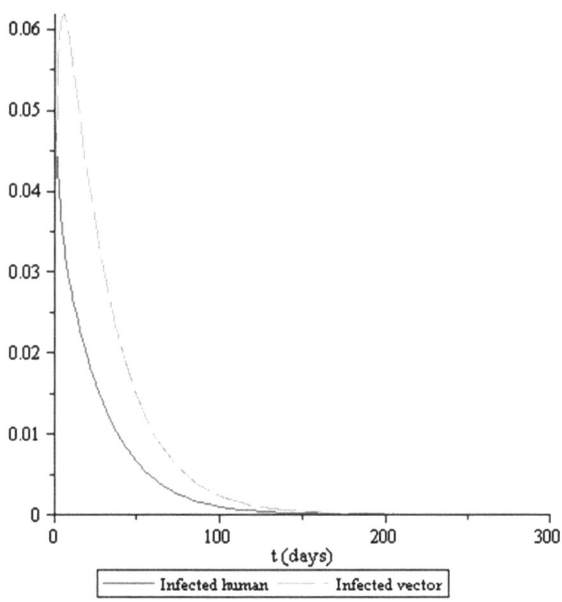

Fig. 5 Numerical solution of I using $A = 500$, $N_H = 10{,}000$

3 Allowance Incubation Period and Without Immunity

In this section, we modified system (2) by adding the assumption that host population can lose its immunity and become susceptible again. We use parameter ϕ to denote the rate of loss of immunity. Because of the possibility that the host population can be reinfected, we assume that there is death rate caused by the disease which is denoted by μ_d. Due to the incubation period, there is the percentage of infected human but incapable of transmitting the disease. If ε is a parameter to describe this condition, then $(1-\varepsilon)$ is the proportion of infected human who are able transmit virus to vector. Figure 6 is the diagram transfer to illustrate model.

Therefore, the mathematical model of dengue disease considered the new assumptions as follow:

$$\begin{aligned}\frac{dS_H(t)}{dt} &= \mu_H N_H - \frac{\beta_H b}{N_H} S_H I_V - \mu_H S_H + \phi R_H \\ \frac{dI_H(t)}{dt} &= \frac{\beta_H b}{N_H} S_H I_V - (\mu_H + \mu_d + \gamma_H) I_H \\ \frac{dR_H(t)}{dt} &= \gamma_H I_H - \mu_H R_H - \phi R_H \\ \frac{dS_V(t)}{dt} &= A - \frac{\beta_V b}{N_H} S_V (1-\varepsilon) I_H - \mu_V S_V \\ \frac{dI_V(t)}{dt} &= \frac{\beta_V b}{N_H} S_V (1-\varepsilon) I_H - \mu_V I_V\end{aligned} \tag{16}$$

with $S_H + I_H + R_H = N_H$ and $S_V + I_V = N_V$. Using the relation $s_H = \frac{S_H}{N_H}$, $i_H = \frac{I_H}{N_H}$, $r_H = \frac{R_H}{N_H}$, $s_V = \frac{S_V}{A/\mu_V}$, $i_V = \frac{I_V}{A/\mu_V}$, $r_H = 1 - (s_H + i_H)$ and $s_V = 1 - i_V$. Then the first equation of system (16) can be written as

$$\frac{ds_H(t)}{dt} = \mu_H (1 - s_H) - \frac{\beta_H b (A/\mu_V)}{N_H} s_H i_V + \phi (1 - s_H - i_H). \tag{17a}$$

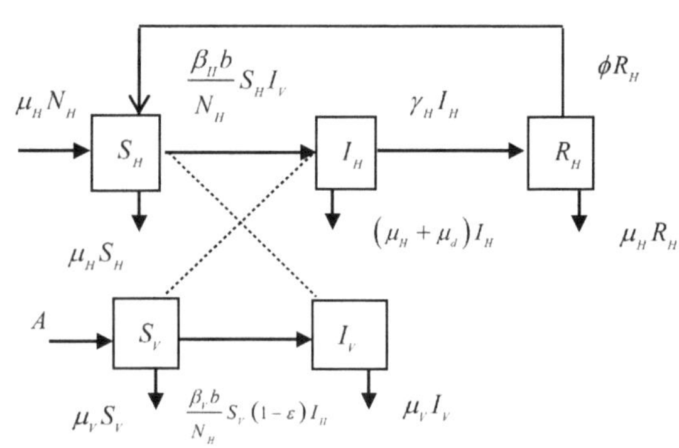

Fig. 6 Diagram transfer for the modified model

The second equation of system (16) can be written as

$$\frac{di_H(t)}{dt} = \frac{\beta_H b\,(A/\mu_V)}{N_H} s_H i_V - (\mu_H + \mu_d + \gamma_H)\, i_H. \tag{17b}$$

For the last equation of system (16), we obtain

$$\frac{di_V}{dt} = \beta_V b\,(1 - i_V)(1 - \varepsilon)\, i_H - \mu_V i_V. \tag{17c}$$

The equilibrium point of system (17) are obtained by letting $\frac{ds_H(t)}{dt} = 0$, $\frac{di_H(t)}{dt} = 0$ and $\frac{di_V(t)}{dt} = 0$. We have disease-free and endemic equilibrium points. Disease-free equilibrium point of system (17) is denoted by $E_2 = \left(\frac{\mu_H}{\phi} + 1, 0, 0\right)$ and the endemic by $E_3 = \left(s_H^*, i_H^*, i_V^*\right)$ with

$$s_H^* = \frac{\phi\left(\frac{\mu_H}{\phi} + 1\right) - \phi i_H^*}{\beta_H b\left(\frac{\frac{\beta_H b}{\mu_V}(1-\varepsilon) i_H^*}{1 + \frac{\beta_H b}{\mu_V}(1-\varepsilon) i_H^*}\right) + \phi}, \quad i_V^* = \frac{\frac{\beta_V b}{\mu_V}(1-\varepsilon)\, i_H^*}{1 + \frac{\beta_V b}{\mu_V}(1-\varepsilon)\, i_H^*}.$$

To determine the basic reproduction number, use the same way like in Sect. 2.2. If $\sigma = \frac{\beta_H \beta_V b^2 \frac{A}{\mu_V}\left(\frac{\mu_H}{\phi}+1\right)}{N_H \mu_V (\mu_H + \mu_d + \gamma_H)}$, then we have basic reproduction number for the second model is $R_0 = \frac{-\sigma\varepsilon}{2} + \sqrt{\left(\frac{\sigma\varepsilon}{2}\right)^2 + \sigma}$.

The Jacobian matrix at $E_2 = \left(\frac{\mu_H}{\phi} + 1, 0, 0\right)$ is

$$J(E_2) = \begin{bmatrix} -\mu_H - \phi & -\phi & -\frac{\beta_H b \frac{A}{\mu_V}}{N_H}\left(\frac{\mu_H}{\phi} + 1\right) \\ 0 & -(\mu_H + \mu_d + \gamma_H) & \frac{\beta_H b \frac{A}{\mu_V}}{N_H}\left(\frac{\mu_H}{\phi} + 1\right) \\ 0 & \beta_V b\,(1 - \varepsilon) & -\mu_V \end{bmatrix}$$

and the Jacobian matrix at $E_3 = \left(s_H^*, i_H^*, i_V^*\right)$ is

$$J(E_3) = \begin{bmatrix} -\frac{\beta_H b}{N_H}\frac{A}{\mu_V} i_V^* - \mu_H - \phi & -\phi & -\frac{\beta_H b}{N_H}\frac{A}{\mu_V} s_H^* \\ \frac{\beta_H b}{N_H}\frac{A}{\mu_V} i_V^* & -(\mu_H + \mu_d + \gamma_H) & \frac{\beta_H b}{N_H}\frac{A}{\mu_V} s_H^* \\ 0 & \beta_V b\left(1 - i_V^*\right)(1 - \varepsilon) & -\beta_V b\,(1 - \varepsilon)\, i_H^* - \mu_V \end{bmatrix}.$$

We have the same result as the first model. If $R_0 < 1$, then all the real part of eigenvalues of $J(E_2)$ are negative, it means the disease-free equilibrium point is asymptotically stable. If $R_0 > 1$, then there is a positive real part of eigenvalue of $J(E_2)$. This condition makes the disease-free equilibrium unstable. Furthermore, if $R_0 > 1$, then the endemic equilibrium point is asymptotically stable.

Using MAPLE, we will show the behavior of model solution. The parameter values are shown in Table 1 and we also use $\mu_d = 0.26$, $\phi = 0.8$, $\varepsilon = 0.3$, and $N_H = 10{,}000$.

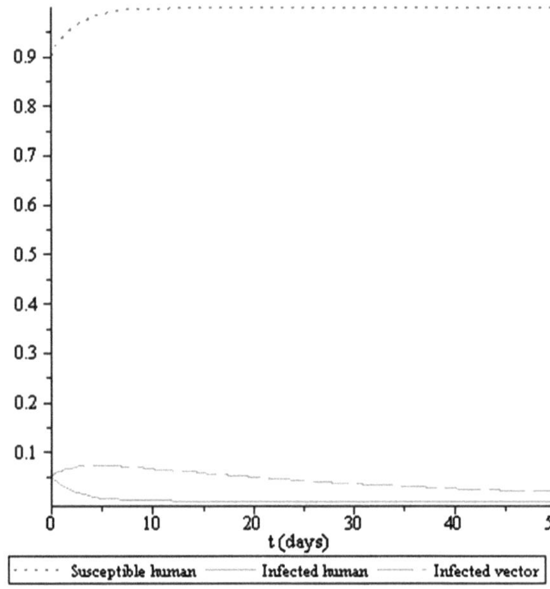

Fig. 7 $A = 10$

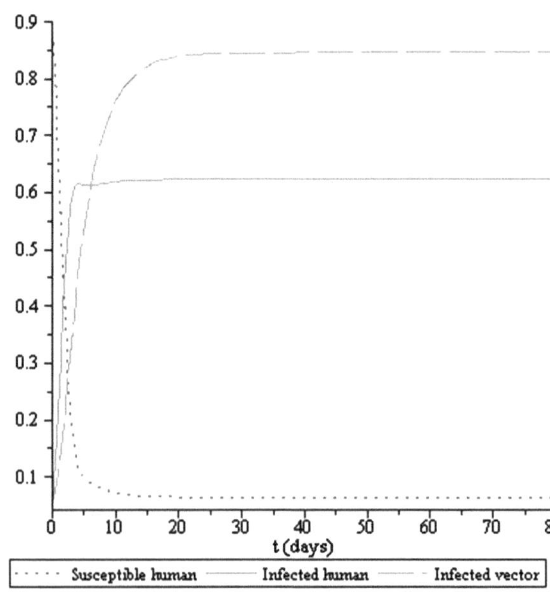

Fig. 8 $A = 5000$

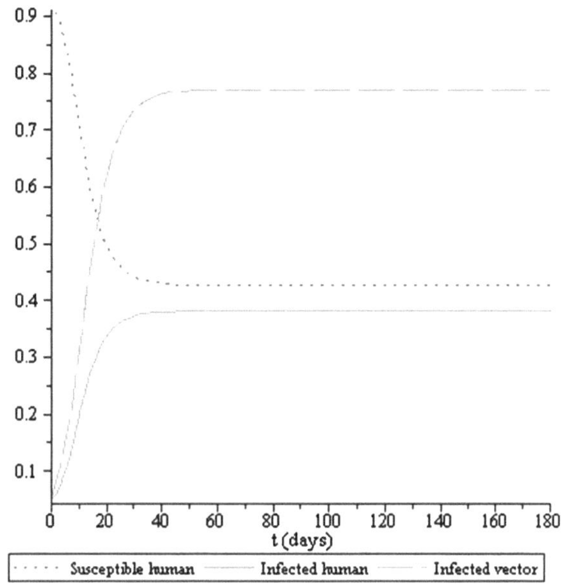

Fig. 9 $A = 500$

It can be seen from Figs. 4, 5, and 7, in order for the solution behavior to approach disease-free equilibrium, it needs smaller value of A. The smaller the value of A given in system (17), the faster the solution to approach disease-free equilibrium.

Using $A = 500$ in Figs. 4 and 5 has not shown the endemic behavior but when it is used in system (17), the solution approach the endemic equilibrium point. The smaller value of A is given, from Figs. 8 and 9, it can be seen the smaller proportion of infected human.

4 Conclusions

We have showed the mathematical model of dengue disease based on host–vector in system (2). The behavior of its solution has been presented analytically and numerically. If the threshold value is greater than one, then the solution will approach the endemic equilibrium point. The more the initial point is given to susceptible human proportion, the more it decreases significantly fast. The more the initial point is given to infectious proportion, the more it increases fast. If the threshold value is less than one, then the solution tends to disease-free equilibrium point.

By adding the assumption to the model, we got system (17). There are disease-free and endemic equilibrium points from system (17). If the threshold value is less than one, it means each infective replaces itself less than one to the new infective, then the disease will die out. On the other hand, if the threshold value is greater than one, then the proportion of S decreases and the infection proportion increases, it means the disease being endemic. We have shown numerically that it just needs a smaller value of A in system (17) than system (2) to be an endemic condition. It is not surprising since that the dengue disease become public health problem.

Acknowledgments We thank the Yogyakarta State University for the funding until this paper can be presented.

References

1. WHO: Dengue and severe dengue. In: http://www.who.int/mediacentre/factsheets/fs117/en/ (2014). Accessed March 2014
2. Rodrigues, H., Monteiro, M., Torres, D., Zinober, A.: Dengue disease basic reproduction number and control. Int. J. Comput. Math. (iFirst) 1–13 (2011)
3. Pusat Data dan Survaliensi Epidemiologi Kemenkes RI: Demam Berdarah Dengue di Indonesia Tahun 1968–2009. Bul. Jendela Epidemiol. **2**, 1–14 (2010)
4. Feng, Z., Hernandez, J.: Competitive exclusion in a vector-host model for the dengue fever. J. Math. Biol. **35**, 523–544 (1997)
5. Soewono, E., Supriatna, A.: A two-dimensional model for the transmission of dengue fever disease. Bul. Malays. Math. Sci. Soc. **24**, 49–57 (2001)
6. Supriatna, A.K., Soewono, E., Gils, S.A.: A two-age-classes dengue transmission model. Math. Biosci. **216**, 114–121 (2008)
7. Aldila, D., Gotz, T., Soewono, E.: An optimal control problem arising from a dengue disease transmission model. Math. Biosci. **242**, 9–16 (2013)
8. Weiss, H.: The SIR model and the foundations of public health. MATerials MATemàtics **17**, 1–17 (2013)
9. Esteva, L., Vargas, C.: Analysis of a dengue disease transmission model. Math. Biosci. **150**, 131–151 (1998)
10. Intergovernmental Panel on Climate Change (IPCC): third assessment report—climate change 2001: impacts, adaptation and vulnerability; Chap. 9 (2001)
11. Ferreira, C., Pinho, S., Esteva, L.: Modelling the dynamics of dengue real epidemics. CNMAC **3**, 89–98 (2010)
12. Driessche, P., Watmough, J.: Reproduction numbers and sub-threshold endemic equilibria for compartmental models of disease transmission. Math. Biosci. **180**, 29–48 (2002)
13. Khan, A., Hassan, M., Imran, M.: Estimating the basic reproduction number for single-strain dengue fever epidemics. Infect. Dis. Poverty **12**(3), 1–17 (2014)
14. Wiggins, S.: Introduction to Applied Nonlinear Dynamical System and Chaos. Springer, New York (2003)

Simple Motion Evasion Differential Game of One Pursuer and One Evader

Idham Arif Alias, Sharifah Anisah Syed Mafdzot and Gafurjan Ibragimov

Abstract We study an evasion differential game of simple motion, involving one pursuer and one evader in the plane $\mathbb{R}^2$. The control functions of players are subjected to geometric constraints. Maximum speed of the pursuer is equal to 1, and maximal speed of the evader is $\alpha > 1$. Control set of the evader is a sector $\mathbf{S}$ whose radius is greater than 1. We say that evasion is possible if the state of the evader does not coincide with that of the pursuer at all times. The problem is to find the conditions of evasion. We obtained conditions that guarantee the evasion regardless of the location of initial position of players.

Keywords Differential game · Geometric constraint · Evasion · Control · Strategy

1 Introduction

The multiple players pursuit–evasion differential games of simple motion with geometric or integral constraints are widely studied in many works (see eg. [1–12]). For the case of many pursuers against one evader, the game is described by the following differential equations:

$$\begin{aligned} &\dot{x}_i = u_i, \quad x_i(0) = x_{i0}, \quad |u_i| \le \rho_i, \quad i = 1, \ldots, m, \\ &\dot{y} = v, \quad y(0) = y_0, \quad |v| \le \sigma, \end{aligned}$$

I.A. Alias · S.A.S. Mafdzot (✉) · G. Ibragimov
Department of Mathematics, Faculty of Science, Universiti Putra Malaysia, 43400 Serdang, Selangor, Malaysia
e-mail: sharifah.anisah@yahoo.com

I.A. Alias
e-mail: idham_aa@upm.edu.my

G. Ibragimov
e-mail: ibragimov@upm.edu.my

A. Kılıçman et al. (eds.), *Recent Advances in Mathematical Sciences*,
DOI 10.1007/978-981-10-0519-0_9

where $x_i, y, u_i, v \in \mathbb{R}^n$, $\rho_i \geq 0$, $\sigma > 0$, $x_{i0} \neq y_0$, and $i = 1, \ldots, m$. Here, pursuit is said to be completed if $x_i(\tau) = y(\tau)$ for some $i \in \{1, \ldots, m\}$ and $\tau \geq 0$, and evasion is said to be possible if $x_i(t) \neq y(t)$ for all $i = 1, \ldots, m$, and $t \geq 0$.

One of the earliest work of evasion differential game in the case of many pursuers versus one evader was by Pshenichnii [11]. In his work, the maximum speed of each player is equal to 1. The necessary and sufficient condition of evasion was obtained where evasion is possible if and only if $y_0 \notin intconv\{x_{10}, \ldots, x_{m0}\}$.

The current paper is closely related to the following work by Chernous'ko [3]. The author examined the evasion differential game of one evader versus many pursuers which is described by the following simple differential equations:

$$\dot{x}_i = u_i, \quad |u_i| \leq kv, \quad 0 < k < 1,$$
$$\dot{y} = v, \quad |v| \leq v.$$

Trajectory of one evader against each pursuer is pictured in the following diagram in Fig. 1 where Q is the position of point P at time t_A; α is the angle between the ray x and segment AQ where $0 \leq \alpha \leq \pi$; R is the current distance QE; φ is the current angle between the segments QE and QA, and s is the arc length of curve AE. The author's strategy is to ensure $EP \geq L$ for all $t \geq t_0$ before the evader moving back to the horizontal line and continue to move away from pursuer horizontally. By the strategy, he proved that evasion is possible.

Also, the work by Ibragimov et al. [5] was devoted on differential game of evasion. The work deals with the evasion differential game of m pursuers $x_1, \ldots, x_m$ and one evader y described by the equations

$$\dot{x}_i = u_i, \quad x_i(0) = x_{i0},$$
$$\dot{y} = v, \quad y(0) = y_0, \quad x_{i0} \neq y_0, \quad i = 1, \ldots, m.$$

The game occurs in the plane and the following conditions:

$$\int_0^\infty |u_i(s)|^2 ds \leq \rho_i^2, \quad i = 1, \ldots, m, \quad \int_0^\infty |v(s)|^2 ds \leq \sigma^2,$$

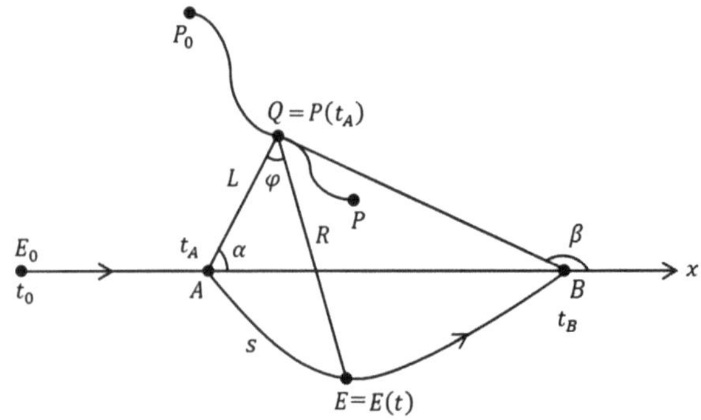

Fig. 1 Trajectory of Evader

were imposed on control functions of the players. The main result of the paper is as follows: if the total resource of pursuers does not exceed that of evader, that is $\rho_1^2 + \rho_2^2 + \ldots \rho_m^2 \leq \sigma^2$, then evasion is possible.

We are to construct a different approach and strategy for evasion problem which is similar to Chernous'ko's [3], for the case of one evader against one pursuer. Like Chernous'ko's problem, evader is moving on a straight line as pursuer is approaching the evader. When the distance between them reaches a particular distance which is greater than zero, the evader will maneuver to avoid the pursuer. The strategy of the maneuver is different from Chernous'ko's where the control set of the evader in our problem is a sector. We also assume the maximal speed of the pursuer is equal to 1 and the maximal speed of the evader is α where $\alpha > 1$. We find a sufficient condition for the evader to escape from the pursuer.

2 Statement of the Problem

We study an evasion differential game of one pursuer P and one evader E with geometric constraints on controls of players described by equations:

$$P : \dot{x} = u, \ x(0) = x_0, \ |u| \leq 1, \tag{1}$$

$$E : \dot{y} = v, \ y(0) = y_0, \ v \in \mathbf{S}, \tag{2}$$

where $x, x_0, u, y, y_0, v \in \mathbb{R}^2$, $x_0 \neq y_0$, u is the control parameter of the pursuer P, v is that of the evader E,

$$\mathbf{S} = \{(v_1, v_2) | \ v_1^2 + v_2^2 \leq \alpha^2, \ |v_1| \leq v_2 \tan\varphi, \ v_2 \geq 0\},$$

$\alpha > 1$ and φ, $0 < \varphi < \frac{\pi}{2}$, is a given angle. Note that $\mathbf{S}$ is a sector with the radius α and central angle 2φ (Fig. 2).

Definition 1 A Borel measurable function $u(t) = (u_1(t), u_2(t))$, $|u(t)| \leq 1$, $t \geq 0$, is called admissible control of the pursuer x.

Definition 2 A Borel measurable function $v(t) = (v_1(t), v_2(t))$, $v(t) \in \mathbf{S}$, $t \geq 0$, is called admissible control of the evader y.

Fig. 2 Sector **S**

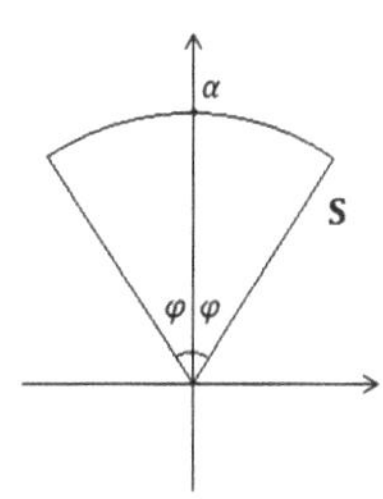

Definition 3 A function

$$V(t, y, x, u),\ \ V : [0, \infty) \times R^2 \times H(0, \rho) \to \mathbf{S},$$

is called strategy of the evader, if for any admissible control $u(t)$ the following initial value problem

$$\dot{x} = u(t),\ \ x(0) = x_0,$$

$$\dot{y} = V(t, y, x, u(t)),\ \ y(0) = y_0,$$

has a unique solution $(x(t), y(t))$, $t \geq 0$, with absolutely continuous components $x(t)$ and $y(t)$.

Definition 4 We say that evasion is possible in the game (1)–(2) if there exists a strategy V of the evader E such that for any admissible control of the pursuer $x(t) \neq y(t)$ for all $t \geq 0$.

Problem

Find sufficient conditions of evasion in the game (1)–(2).

The condition $v(t) \in \mathbf{S}$, $t \geq 0$, implies that state of the evader $y(t)$ belongs to the sector $\mathbf{S}_1 = \{y_0 + ta \mid a \in \mathbf{S}, t \geq 0\}$. Initial position of the pursuer may be in $\mathbf{S}_1$ as well as outside $\mathbf{S}_1$. In process of pursuit, the pursuer can move throughout the plane. In this regard, the pursuer have advantage. However, the evader has advantage in speed, since $\alpha > 1$.

3 Main Result

The main result of the paper is the following statement.

Theorem 1 *If* $\alpha \cos \varphi_0 \geq 1$ *and* $\alpha \sin \varphi_0 > 1$ *at some* $0 < \varphi_0 \leq \varphi$, *then evasion is possible in the game* (1)–(2).

Proof Choose any number $a, 0 < a < \min |x_0 - y_0|$. We construct a strategy for the evader as follows:

$$v(t) = \begin{cases} (0, \alpha)\,,\ if\ 0 \leq t < \tau, \\ \left(\pm(c + |u_1(t)|), \sqrt{\alpha^2 - (c + |u_1(t)|)^2}\right), \\ \qquad\qquad\qquad\qquad if\ \tau \leq t \leq t_1, \\ (0, \alpha)\,,\ if\ t > t_1, \end{cases} \tag{3}$$

where τ is the first time when $|y(\tau) - x(\tau)| = a$,

$$t_1 = \tau + \frac{a}{\sqrt{(\alpha - 1)^2 - c^2}},\quad c = \alpha \sin \varphi_0 - 1.$$

Note that such a time τ may not exist. If so, then we let $v(t) = (0, \alpha)$ for all $t \geq 0$. In (3), $\pm$ means $v_1(t) = c + |u_1(t)|$, if $x_1(\tau) \leq y_1(\tau)$ and $v_1(t) = -(c + |u_1(t)|)$, if $x_1(\tau) \geq y_1(\tau)$, where $x(t) = (x_1(t), x_2(t))$, $y(t) = (y_1(t), y_2(t))$.

Estimate $|y(t) - x(t)|$, $\tau \leq t \leq t_1$. We have

$$
\begin{aligned}
|y(t) - x(t)| &= \left| y(\tau) + \int_\tau^t v(s)ds - \left(x(\tau) + \int_\tau^t u(s)ds \right) \right| \\
&\geq |y(\tau) - x(\tau)| - \left| \int_\tau^t v(s)ds \right| - \left| \int_\tau^t u(s)ds \right| \\
&\geq a - (t - \tau)(\alpha + 1).
\end{aligned}
$$

Without loss of generality, we assume that $y_1(\tau) \geq x_1(\tau)$, and hence $v_1(t) = c + |u_1(t)|$, $\tau \leq t \leq t_1$. Then, on the other hand, for the points $x(t)$ and $y(t)$ we have

$$
\begin{aligned}
|y(t) - x(t)| &\geq y_1(t) - x_1(t) \\
&= y_1(\tau) + \int_\tau^t v_1(s)ds - \left(x_1(\tau) + \int_\tau^t u_1(s)ds \right) \\
&= y_1(\tau) - x_1(\tau) + \int_\tau^t (v_1(s) - u_1(s))\,ds \\
&= y_1(\tau) - x_1(\tau) + \int_\tau^t (c + |u_1(s)| - u_1(s))\,ds \\
&\geq c \cdot (t - \tau)
\end{aligned}
$$

Thus, $|y(t) - x(t)| \geq f(t)$, where

$$
f(t) = \max\{a - (t - \tau)(\alpha + 1),\ c(t - \tau)\}.
$$

Note that the function $f(t)$ has only one minimum on $[\tau, t_1]$, since the first function in the max decreases, whereas the second function increases. The function $f(t)$ takes its minimum at

$$
t_* = \tau + \frac{a}{\alpha(1 + \sin\varphi_0)} \in [\tau, t_1].
$$

We have

$$
\begin{aligned}
|y(t) - x(t)| &\geq c \cdot (t_* - \tau) \\
&= c \cdot \frac{a}{\alpha(1 + \sin\varphi_0)} \\
&\geq \frac{c}{2\alpha} \cdot a, \qquad \tau \leq t \leq t_1.
\end{aligned}
$$

In particular, at the time t_1

$$|y(t_1) - x(t_1)| \geq \frac{c}{2\alpha} \cdot a. \tag{4}$$

Moreover, at the time $t = t_1$ the pursuer cannot be above the horizontal line $y = y_2(t_1)$ of the xy-plane. Indeed,

$$\begin{aligned} y_2(t_1) - x_2(t_1) &= y_2(\tau) + \int_\tau^{t_1} v_2(t)dt - \left(x_2(\tau) + \int_\tau^{t_1} u_2(t)dt\right) \\ &= y_2(\tau) - x_2(\tau) + \int_\tau^{t_1} (v_2(t) - u_2(t))\, dt \\ &\geq -a + \int_\tau^{t_1} \left(\sqrt{\alpha^2 - (c + |u_1(t)|)^2} - \sqrt{1 - |u_1(t)|^2}\right) dt \end{aligned} \tag{5}$$

It is not difficult to show that

$$\sqrt{\alpha^2 - (c + |u_1(t)|)^2} - \sqrt{1 - |u_1(t)|^2} \geq \sqrt{(\alpha - 1)^2 - c^2}.$$

Then right-hand side of (5) can be estimated from below by

$$-a + \int_\tau^{t_1} \sqrt{(\alpha - 1)^2 - c^2 dt} = -a + (t_1 - \tau)\sqrt{(\alpha - 1)^2 - c^2} = 0$$

Thus, $y_2(t_1) \geq x_2(t_1)$.

Next, according to (3) $v(t) = (0, \alpha)$, $t > t_1$. We estimate $|x(t) - y(t)|$ at $t > t_1$. Since

$$y_1(t) - x_1(t) = y_1(t_1) - x_1(t_1) - \int_{t_1}^{t} u_1(s)ds$$

and

$$y_2(t) - x_2(t) \geq y_2(t_1) - x_2(t_1) + \alpha(t - t_1) - \int_{t_1}^{t} |u_2(s)|ds \geq 0.$$

Last inequality because of $y_2(t_1) \geq x_2(t_1)$ and $|u_2(t)| \leq 1$.
Therefore

$$\begin{aligned} |x(t) - y(t)| \geq &\left(\left(y_1(t_1) - x_1(t_1) - \int_{t_1}^{t} u_1(s)ds\right)^2\right. \\ &\left. + \left(y_2(t_1) - x_2(t_1) + \alpha(t - t_1) - \int_{t_1}^{t} |u_2(s)|ds\right)^2\right)^{1/2}. \end{aligned}$$

Using the triangle inequality, we can estimate right-hand side of this inequality from below by

$$\sqrt{|y_1(t_1) - x_1(t_1)|^2 + (y_2(t_1) - x_2(t_1) + \alpha(t - t_1))^2}$$
$$-\sqrt{\left(\int_{t_1}^{t} u_1(s)ds\right)^2 + \left(\int_{t_1}^{t} |u_2(s)|ds\right)^2}.$$

By the Cauchy–Schwarz inequality we obtain

$$\begin{aligned}\left(\int_{t_1}^{t} u_1(s)ds\right)^2 + \left(\int_{t_1}^{t} |u_2(s)|ds\right)^2 &\le \left(\int_{t_1}^{t} 1\cdot|u_1(s)|ds\right)^2 + \left(\int_{t_1}^{t} 1\cdot|u_2(s)|ds\right)^2\\ &\le \int_{t_1}^{t} 1^2 ds \int_{t_1}^{t} u_1^2(s)ds + \int_{t_1}^{t} 1^2 ds \int_{t_1}^{t} u_2^2(s)ds\\ &= (t - t_1)\int_{t_1}^{t} \left(u_1^2(s) + u_2^2(s)\right) ds\\ &\le (t - t_1)^2.\end{aligned}$$

Therefore,

$$\begin{aligned}|x(t) - y(t)| &\ge \left((y_1(t_1) - x_1(t_1))^2 + (y_2(t_1) - x_2(t_1))^2 + (\alpha(t - t_1))^2\right)^{1/2} - (t - t_1)\\ &\ge \sqrt{\left(\frac{ac}{2\alpha}\right)^2 + \alpha^2(t - t_1)^2} - (t - t_1) \qquad (6)\end{aligned}$$

since $|y(t_1) - x(t_1)|^2 \ge (\frac{ac}{2\alpha})^2$.

To estimate the expression (6) from below, we let $\dfrac{ac}{2\alpha} = b$, $(t - t_1) = x$ and consider the function

$$g(x) = \sqrt{b^2 + \alpha^2 x^2} - x, \quad x \ge 0.$$

This function takes its minimum at $x = x_* = \dfrac{b}{\alpha\sqrt{\alpha^2 - 1}}$. Thus

$$\begin{aligned}|x(t) - y(t)| &\ge g(x_*)\\ &= \frac{\sqrt{\alpha^2 - 1}}{\alpha} b = \frac{\sqrt{\alpha^2 - 1}}{2\alpha^2} ac, \quad t \ge t_1.\end{aligned}$$

Next, we estimate $y_2(t) - x_2(t)$ for $t \ge \tau' = \tau + ra$ where

$$r = \frac{1}{\sqrt{(\alpha - 1)^2 - c^2}} + \frac{c}{2\alpha^2\sqrt{\alpha^2 - 1}}.$$

Since $y_2(t_1) \geq x_2(t_1)$,

$$\begin{aligned} y_2(t) - x_2(t) &= y_2(t_1) + \int_{t_1}^{t} v_2(s)ds - x_2(t_1) - \int_{t_1}^{t} u_2(s)ds \\ &\geq y_2(t_1) - x_2(t_1) + \int_{t_1}^{t} \alpha\, ds - \int_{t_1}^{t} 1\, ds \\ &\geq (\alpha - 1)(t - t_1) \\ &\geq (\alpha - 1)(\tau' - t_1) \\ &= \frac{(\alpha - 1)c}{2\alpha^2\sqrt{\alpha^2 - 1}} \cdot a \\ &= ap, \end{aligned}$$

where $p = \dfrac{(\alpha - 1)c}{2\alpha^2\sqrt{\alpha^2 - 1}}$. □

Thus, we can conclude that

(1) $|x(t) - y(t)| \geq a$ for $0 \leq t \leq \tau$,
(2) $|x(t) - y(t)| \geq \frac{c}{2\alpha} \cdot a$ for $\tau \leq t \leq t_1$,
(3) $y_2(t) > x_2(t)$, for $t \geq t_1$,
(4) $|x(t) - y(t)| \geq \dfrac{\sqrt{\alpha^2 - 1}}{2\alpha^2} ac$ for $t \geq t_1$,
(5) $y_2(t) - x_2(t) \geq ap$ for $t \geq \tau'$.

In particular, $y(t) \neq x(t),\ t > 0$.

Hence, at the time t_1 the evader will be above the horizontal line where is the pursuer x. Thus, at the time t_1 the pursuer x became "behind" the evader. Since $v_2(t) \geq \alpha \cos \varphi_0 \geq 1 \geq u_2(t)$, then $y_2(t) > x_2(t)$, for all $t \geq t_1$.

If a-approach occurs with the pursuer x at a time τ, then the evader uses maneuver on $[\tau, t_1]$ which ensures the inequality $y_2(t_1) \geq x_2(t_1)$. Further, the strategy of the evader guarantees him the inequality $y_2(t) > x_2(t)$ for all $t \geq t_1$.

4 Conclusion

A simple motion evasion differential game of one pursuer and one evader whose control set is a sector has been considered in the plane. If $\alpha \cos \varphi_0 \geq 1$ and $\alpha \sin \varphi_0 > 1$, then evasion from one pursuer has been presented. The strategy for the evader was constructed as well. Moreover, distances between the pursuer and evader have been estimated.

Acknowledgments The authors would like to thank the referees for their helpful comments which improved and shortened arguments of the present paper. This research was partially supported by the Research Grant (RUGS) of the Universiti Putra Malaysia, No. 05-02-12-1868RU.

References

1. Azamov, A.: On a problem of escape along a prescribed curve. J. Appl. Math. Mech. **46**(4), 553–555 (1982)
2. Azamov, A.A.: Fundamentals of Theory of Discrete Games. Niso Poligraf, Tashkent, Uzbekistan (2011)
3. Chernous'ko, F.L.: A problem of evasion from many pursuers. Prikladnaya Matematika I Mekhanika **40**(1), 14–24 (1976)
4. Chernousko, F.L., Zak, V.L.: On Differential Games of Evasion from Many Pursuers. J. Optim. Theory Appl. **46**(4), 461–470 (1985)
5. Ibragimov, G.I., Salimi, M., Amini, M.: Evasion from many pursuers in simple motion differential game with integral constraints. Eur. J. Oper. Res. **218**(2), 505–511 (2012)
6. Isaacs, R.: Control Optim. Differential Games: A Mathematical Theory with Applications to Warfare and Pursuit. Wiley, New York (1965)
7. Ivanov, R.P.: Simple pursuit-evasion on a compact convex set. Doklady Akademii Nauk SSSR **254**(6), 1318–1321 (1980)
8. Kuchkarov, ASh: Solution of simple pursuit-evasion problem when evader moves on a given curve. Int. Game Theory Rev. **12**(3), 223–238 (2010)
9. Petrov, N.N., Shchelchkov, K.A.: On the interrelation of two linear stationary evasion problems with many evaders. Vestn. Udmurtsk. Univ. Mat. Mekh. Komp. Nauki **3**, 52–58 (2014)
10. Pontryagin, L.S.: Selected Works. Nauka, Moscow (1988)
11. Pshenichnii, B.N.: Simple pursuit by several objects. Cybern. Syst. Anal. **12**(3), 145–146 (1976)
12. Zak, V.L.: On a problem of evading many pursuers. J. Appl. Math. **43**(3), 456–465 (1978)

Author Index

A. Kılıçman et al. (eds.), *Recent Advances in Mathematical Sciences*,
DOI 10.1007/978-981-10-0519-0

MIX
Papier aus verantwortungsvollen Quellen
Paper from responsible sources
FSC® C105338
FSC
www.fsc.org

If you have any concerns about our products,
you can contact us on
ProductSafety@springernature.com

In case Publisher is established outside the EU,
the EU authorized representative is:
Springer Nature Customer Service Center GmbH
Europaplatz 3, 69115 Heidelberg, Germany

Printed by Libri Plureos GmbH
in Hamburg, Germany